Frontiers in Hybrid Vehicles Powertrain

Frontiers in Hybrid Vehicles Powertrain

Editors

Marco Cammalleri
Vincenzo Di Dio
Antonella Castellano

MDPI

Basel • Beijing • Wuhan • Barcelona • Belgrade • Novi Sad • Cluj • Manchester

Editors
Marco Cammalleri
University of Palermo
Palermo, Italy

Vincenzo Di Dio
University of Palermo
Palermo, Italy

Antonella Castellano
University of Palermo
Palermo, Italy

Editorial Office
MDPI
St. Alban-Anlage 66
4052 Basel, Switzerland

This is a reprint of articles from the Special Issue published online in the open access journal *Applied Sciences* (ISSN 2076-3417) (available at: https://www.mdpi.com/journal/applsci/special_issues/Hybrid_Vehicles_Powertrain).

For citation purposes, cite each article independently as indicated on the article page online and as indicated below:

Lastname, A.A.; Lastname, B.B. Article Title. *Journal Name* **Year**, *Volume Number*, Page Range.

ISBN 978-3-0365-8628-1 (Hbk)
ISBN 978-3-0365-8629-8 (PDF)
doi.org/10.3390/books978-3-0365-8629-8

Contents

Contents

About the Editors

Marco Cammalleri

Marco Cammalleri is a Full Professor of Applied Mechanics at the Department of Engineering of the University of Palermo (Italy). He graduated in Mechanical Engineering and has spent his entire academic career at the University of Palermo where he held several positions of responsibility. He is currently the Director of the Historic Museum of Engines and Mechanisms of his University. His research activity has involved several topics, such as V-belt mechanics, mechanics of vibrations, rotor dynamics, functional design and control of machines for linear friction welding processes, dynamics of solenoid injectors, food engineering, and analysis of devices of historical interest; however, his activity is mainly focused on Power-Split Continuously Variable Transmissions (PS-CVTs) for hybrid electric vehicles. Among his scientific and professional memberships, he is a member of Italian M.O. of the International Federation for the Promotion of Mechanism and Machine Science (IFToMM).

Vincenzo Di Dio

Vincenzo Di Dio is currently an Associate Professor of Converters, Electrical Machines, and Drives at the Department of Engineering of the University of Palermo (Italy). He is a member of many research groups for national and international scientific cooperation among the University of Palermo and others. His main research interests include mathematical models of electrical machines, drive system control and diagnostics, renewable energies and energy management. He is a member of the Italian Standardization Body (CEI - Comitato Elettrotecnico Italiano) and scientific supervisor of research projects in the field of smart cities and generation of electrical energy from renewable resources.

Antonella Castellano

Antonella Castellano received her Master of Science in Mechanical Engineering cum laude from the University of Palermo (Italy) in 2020. She is currently a Ph.D. student at the same institution. Her research deals with hybrid electric powertrain, with a special focus on the power-split hybrid system. The main research themes cover both the mechanical design and the control problem for hybrid electric powertrain, applied mostly - but not only - to the automotive sector. She has been a visiting student at the Centre for Automotive Engineering of the University of Surrey (UK) to develop an energy management strategy based on model predictive control for power-split hybrid electric vehicles.

Preface

The increasing concern regarding environmental issues has led to the adoption of stringent regulations worldwide to facilitate the urgent move towards green mobility and sustainable transportation. In this regard, electrified powertrains are bound to replace conventional thermal engines to reduce greenhouse gases and pollutant emissions.

Although full electric vehicles succeed in reducing onboard tank-to-wheel emissions, they are still affected by some major issues, such as range anxiety, battery lifetime, high cost, and limited recharging infrastructures. Moreover, the low specific energy of a pure electric powertrain prevents its final establishment beyond light- and medium-duty ground vehicles. On the contrary, the synergy between an internal combustion engine and an electric unit on hybrid electric vehicles (HEVs) overcomes the main battery limitations and fulfills even the most demanding energetic requirements, ranging from light- and heavy-duty vehicles to agricultural machinery, vessels, and aircraft.

As a result, HEVs significantly reduce fuel consumption and emissions while maintaining high vehicle performance and driving comfort, becoming the optimal sustainable solution in the short term. Nonetheless, the full potential of HEVs can only be exploited using a multidisciplinary approach to design the mechanical and electrical equipment and implement an optimal energy management strategy to pursue the best trade-off between fuel saving and performance.

The main goal of this Special Issue, "Frontiers in Hybrid Vehicles Powertrain", has been to gather original contributions from several scientific fields to provide a broad perspective on the current issues and research trends on the hybrid electric powertrain. In this respect, this book collects nine research articles covering efficiency-oriented control strategies and novel mathematical tools for power-split hybrid transmissions to reduce pollutant emissions, improve powertrain overall efficiency, and enhance battery life. Moreover, the adoption of the hybrid electric powertrain on agricultural machinery and air–ground vehicles is analyzed, along with the integration of the latest advancements towards zero-emission mobility regarding hydrogen-powered engines and renewable energy sources. Overall, the Special Issue outlines the major unsolved challenges for the final widespread establishment of HEVs.

Thanks are due to all the authors who contributed to this Special Issue, to the reviewers who improved the quality of the work, and to the editorial team of the *Applied Sciences* journal and MDPI Book staff.

Marco Cammalleri, Vincenzo Di Dio, and Antonella Castellano
Editors

applied sciences

MDPI

Editorial

Special Issue on Frontiers in Hybrid Vehicles Powertrain

Marco Cammalleri *, Vincenzo Di Dio and Antonella Castellano

Department of Engineering, University of Palermo, 90128 Palermo, Italy; vincenzo.didio@unipa.it (V.D.D.); antonella.castellano@unipa.it (A.C.)
* Correspondence: marco.cammalleri@unipa.it

1. Introduction

The urgent need to reduce greenhouse gases and toxic emissions is fostering a rapid shift towards more sustainable transport. In this respect, hybrid electric vehicles (HEVs) emerge as a sustainable solution to increasingly adopt on a large scale very shortly. In HEVs, the internal combustion engine (ICE) efficiency is significantly improved by cooperating with an electric unit, consisting of one or more electric machines and a battery pack, that supports the engine during transient operations and enables regenerative braking. As a result, a significant reduction in fuel consumption and emissions can be achieved while maintaining high vehicle performance and driving comfort. A wide variety of hybrid electric powertrains is currently available on the market, ranging from various degrees of electrification, i.e., mild, full, and plug-in hybrid, and different driveline architectures, i.e., series, parallel, and power-split hybrid [1]. The main challenges of HEVs lie in the powertrain design, not only in terms of thermal, mechanical, and electrical components but also of energy management strategy. Consequently, effective research efforts in this field should integrate contributions and knowledge from several scientific areas, such as mechanical, electrical, controls, chemical, and energetic. In this respect, the Special Issue succeeded in collecting nine research articles coming from the sectors mentioned above, providing a broad perspective on open challenges for HEVs.

2. Hybrid Electric Vehicles: A Multidisciplinary Challenge

Although the hybrid electric powertrain is potentially more efficient than the traditional ICE-based one, implementing an effective energy management strategy (EMS) is crucial to achieve an actual reduction in fuel consumption and emissions without compromising vehicle performance. Several EMSs have been proposed in the literature [2], but this is still a major research issue. In this respect, Dimauro et al. [3] investigated the transient dynamic response of a parallel hybrid electric truck. The parallel hybrid layout includes only one electric machine that can boost the tractive power during the acceleration phases or recover energy during the deceleration phases, while a direct mechanical link between the ICE and the wheel is maintained. The authors proposed a control strategy based on torque allocation to split power between the engine and the electric motor. The results showed a satisfying trade-off between vehicle drivability and NOx emissions, reduced by limiting the ICE transient operations.

Another critical issue in HEVs concerns the battery lifetime, which should be considered in the development of the EMS. Anselma et al. [4] introduced an optimization-driven methodology to tune the parameters of thermal and energy on-board rule-based control approaches of a parallel through-the-road plug-in HEV. They used particle swarm optimization to minimize energy consumption and battery degradation by considering various ambient temperatures, driving conditions, payload conditions, and cabin conditioning system states. The obtained results suggested that pure electric operations should be reduced as the ambient temperature progressively increases beyond 30 °C, proving that temperature significantly affects the battery state of health and, consequently, the fuel

Citation: Cammalleri, M.; Di Dio, V.; Castellano, A. Special Issue on Frontiers in Hybrid Vehicles Powertrain. *Appl. Sci.* **2023**, *13*, 6367. https://doi.org/10.3390/app13116367

Received: 5 May 2023
Accepted: 11 May 2023
Published: 23 May 2023

consumption. Preserving the battery life was also pursued by Li et al. [5], who proposed an EMS for a series hybrid electric air-ground vehicles that can both run on land and fly. Series hybrid includes an electric generator directly linked to the ICE, which is mechanically uncoupled from the wheels, and another electric machine providing the power coming from the generator or the batteries for the propulsion or performing regenerative braking tasks. The EMS presented by the authors involves Pontryagin's minimum principle-based model predictive control framework where speed information from intelligent network technology predicted the reference trajectory for the battery state of charge. Compared with more traditional EMS, fuel efficiency was improved and battery temperature was reduced.

As the latter contribution suggested, a hybrid powertrain can be effectively adopted in those applications where the complete disposal of the ICE in favor of a pure electric powertrain appears less practicable. In this regard, Mocera and Martini [6] developed a power-split hybrid electric layout for an orchard tractor. The power-split hybrid architecture merges the advantages of both series and parallel hybrid and overcomes their drawbacks. Indeed, the ICE is kinematically decoupled from the wheels, such as in the series hybrid. However, some of its power can still be delivered for vehicle propulsion without any intermediate electric conversion, such as in the parallel hybrid. Consequently, the ICE could always operate within the most efficient operation and a significant downsizing of both the engine and the electric unit can be realized. The authors presented two control strategies: a charge depleting mode to use full power for the most power-intensive scenarios and a charge sustaining mode to optimize efficiency and battery use throughout the whole working day. Promising results in terms of peak power capabilities and fuel savings were obtained.

In the power-split hybrid architecture, the ICE, two electric machines and the output shaft are coupled by a power-split unit consisting of one or more planetary gear sets, as well as ordinary gear sets. Furthermore, embedding a clutch system enables multimode operations to increase powertrain adaptability in various road conditions. Nonetheless, the high constructive complexity of power-split transmissions requires proper mathematical tools for the analysis and the design. Castellano and Cammalleri [7], taking the Voltec multimode power-split transmission as a case study, described a procedure to evaluate a global efficiency map in various battery scenarios that can be exploited to extract data to implement a real-time EMS. The contribution is based on a universal parametric approach that can be applied to any power-split transmission. Besides the ICE and electric machines efficiency, it also considers the transmission meshing losses, which are rather challenging to evaluate, especially for multimode architecture with multiple planetary gearing, and thus often neglected. In the same regard, Esmail et al. [8] proposed general and reliable formulas to compute the mechanical efficiency of a planetary gear set. The new formulation exploits the concept of potential or virtual power and highlights the dependency between kinematics and efficiency. Mantriota et al. [9] evaluated the power flows and the efficiency of a power-split transmission with two planetary gear sets, showing that adopting this solution would be more advisable if included in a multimode framework. The study aimed to provide designers with the advantages and disadvantages of implementing this topology within a complex compound powertrain, paving the path for developing an optimization tool that identifies the most efficient configuration for each working condition.

The last two contributions foresee the embedding on HEVs of some of the most cutting-edge advancements towards zero-emission mobility. Beccari [10] proposed the implementation of a hydrogen-fueled engine in a power-split hybrid vehicle. By supposing that the ICE always operates at its most efficient point, numeric simulations were performed in order to compare the gasoline-fueled engine with the hydrogen-fueled one in terms of the thermal efficiency and total energy consumed during a driving cycle. The results showed a mean engine efficiency increase of around 17% and an energy consumption reduction of around 15%. Ramshanker et al. [11] investigated greenhouse gas emissions when a floating solar photovoltaic system is used to recharge the vehicle. In this case, land exploitation is reduced and the decrease in evaporation saves water. The results showed a promising reduction in CO_2 emissions.

3. Research Trends for HEVs

The contributions collected in this Special Issue effectively reflect the challenges that prevent the desirable widespread uptake of HEVs. Prediction-based real-time energy management strategies are crucial to reduce fuel consumption and emissions while maintaining high vehicle performance. This is true especially for the most promising layout, the power-split hybrid, whose higher constructive complexity enables more efficient operations but requires proper mathematical tools for an effective control strategy and a conscious design. Moreover, HEVs share similar battery issues with pure-electric vehicles related to battery performance optimization and battery life safeguard. Lastly, if the transition towards pure-electric ground mobility seems to become a reality soon, there are other fields (i.e., agricultural machinery, aircraft, and air-ground vehicles) where the replacement of the thermal unit looks less forthcoming. More efforts towards the development of hybrid electric powertrain beyond light-duty vehicles are advisable, along with the integration of the latest innovations in green fuel and renewable energy.

Acknowledgments: Thanks are due to all the authors who contributed to this Special Issue, making it successful and providing the scientific community with a multidisciplinary perspective on hybrid electric vehicles. Special thanks to all the reviewers who enhanced the quality of each contribution with their valuable comments and suggestions. Last but not least, we would like to thank the editorial team of *Applied Sciences* for their constant and dedicated support.

Conflicts of Interest: The authors declare no conflict of interest.

References

1. Zhuang, W.; Li, S.; Zhang, X.; Kum, D.; Song, Z.; Yin, G.; Ju, F. A Survey of Powertrain Configuration Studies on Hybrid Electric Vehicles. *Appl. Energy* **2020**, *262*, 114553. [CrossRef]
2. Zhang, F.; Wang, L.; Coskun, S.; Pang, H.; Cui, Y.; Xi, J. Energy Management Strategies for Hybrid Electric Vehicles: Review, Classification, Comparison, and Outlook. *Energies* **2020**, *13*, 3352. [CrossRef]
3. Dimauro, L.; Tota, A.; Galvagno, E.; Velardocchia, M. Torque Allocation of Hybrid Electric Trucks for Drivability and Transient Emissions Reduction. *Appl. Sci.* **2023**, *13*, 3704. [CrossRef]
4. Anselma, P.G.; Del Prete, M.; Belingardi, G. Battery High Temperature Sensitive Optimization-Based Calibration of Energy and Thermal Management for a Parallel-through-the-Road Plug-in Hybrid Electric Vehicle. *Appl. Sci.* **2021**, *11*, 8593. [CrossRef]
5. Li, Z.; Jiao, X.; Zha, M.; Yang, C.; Yang, L. Predictive Energy Management Strategy for Hybrid Electric Air-Ground Vehicle Considering Battery Thermal Dynamics. *Appl. Sci.* **2023**, *13*, 3032. [CrossRef]
6. Mocera, F.; Martini, V. Numerical Performance Investigation of a Hybrid ECVT Specialized Agricultural Tractor. *Appl. Sci.* **2022**, *12*, 2438. [CrossRef]
7. Castellano, A.; Cammalleri, M. Power Losses Minimization for Optimal Operating Maps in Power-Split Hevs: A Case Study on the Chevrolet Volt. *Appl. Sci.* **2021**, *11*, 7779. [CrossRef]
8. Esmail, E.L.; Pennestrì, E.; Cirelli, M. Power-Flow and Mechanical Efficiency Computation in Two-Degrees-of-Freedom Planetary Gear Units: New Compact Formulas. *Appl. Sci.* **2021**, *11*, 5991. [CrossRef]
9. Mantriota, G.; Reina, G.; Ugenti, A. Performance Evaluation of a Compound Power-Split CVT for Hybrid Powertrains. *Appl. Sci.* **2021**, *11*, 8749. [CrossRef]
10. Beccari, S. On the Use of a Hydrogen-Fueled Engine in a Hybrid Electric Vehicle. *Appl. Sci.* **2022**, *12*, 12749. [CrossRef]
11. Ramshanker, A.; Chakraborty, S.; Elangovan, D.; Kotb, H.; Aboras, K.M.; Giri, N.C.; Agyekum, E.B. CO_2 Emission Analysis for Different Types of Electric Vehicles When Charged from Floating Solar Photovoltaic Systems. *Appl. Sci.* **2022**, *12*, 12552. [CrossRef]

applied sciences

MDPI

Article

Torque Allocation of Hybrid Electric Trucks for Drivability and Transient Emissions Reduction

Luca Dimauro, Antonio Tota, Enrico Galvagno and Mauro Velardocchia *

Department of Mechanical and Aerospace Engineering, Politecnico di Torino, Corso Duca Degli Abruzzi 24, 10129 Torino, Italy
* Correspondence: mauro.velardocchia@polito.it

Abstract: This paper aims at investigating powertrain behaviour, especially in transient dynamic responses, using a nonlinear truck vehicle dynamic model with a parallel hybrid configuration. A power split control was designed to achieve the desired drivability performance, with a focus on NOx emissions. The controller was characterized by high-level model-based logic used to elaborate the total powertrain torque required, and a low-level allocation strategy for splitting power between the engine and the electric motor. The final task was to enhance vehicle drivability based on driver requests, with the goal of reducing—in a hybrid configuration—transient diesel engine emissions when compared to a conventional pure thermal engine powertrain. Different parameters were investigated for the assessment of powertrain performance, in terms of external input disturbance rejection and NOx emissions reduction. The investigation of torque allocation performance was limited to the simulation of a Tip-in manoeuvre, which showed a satisfying trade-off between vehicle drivability and transient emissions.

Keywords: parallel hybrid truck vehicle model; NOx emissions; torque allocation control logic; phlegmatization; drivability and emission trade-off

Citation: Dimauro, L.; Tota, A.; Galvagno, E.; Velardocchia, M. Torque Allocation of Hybrid Electric Trucks for Drivability and Transient Emissions Reduction. *Appl. Sci.* **2023**, *13*, 3704. https://doi.org/10.3390/app13063704

Academic Editors: Vincenzo Di Dio, Antonella Castellano and Marco Cammalleri

Received: 22 February 2023
Revised: 8 March 2023
Accepted: 9 March 2023
Published: 14 March 2023

1. Introduction

In recent years, an ever-increasing number of researchers have focused their efforts on the substantial challenge of powertrain electrification in an attempt to deal with new government requirements that aim at reducing pollutant emissions and fuel consumption [1]. In this context, the Research and Development (R&D) departments of several automobile companies are engaged in the development of innovative powertrain solutions to replace vehicles with fuel motorization. A slow and progressive transition from internal combustion engine vehicles (ICEVs) to zero emissions vehicles (ZEVs), battery electric vehicles (BEVs), and fuel cell electric vehicles (FCEVs) is occurring, along with the development of hybrid electric vehicles (HEVs) and plug-in hybrid electric vehicles (PHEVs).

As proposed in [2], several possible topological architectures can be implemented in HEVs, where two or more sources of power/energy are combined to achieve the required power necessary to propel the vehicle. HEV vehicles combine the ICE power with an electric traction motor, which is powered by an energy storage device generally known as battery pack. The most common configurations adopted to couple different power sources can be summarized as parallel hybrid [3], series hybrid [4] or combined series–parallel architectures, with a mechanical power-split [5–8] or innovative magnetic split [9]. Different control strategies have been investigated and proposed to achieve optimal energy management [10] between different power sources. HEVs combine the advantages of electric motor drives—having quick acceleration—with good ICE performance at constant speeds. On the other hand, vehicle drivability [11] is highly affected, as during rapid torque transients—due to drivers' abrupt accelerations—noise and vibrations [12] can occur during engine start/stop conditions [13] in mild hybrid electric vehicles.

Besides this aspect, sudden driver torque requests can generate an increase in NOx emissions, as investigated in [14] during tip-in/tip-out manoeuvres. Hence, control strategies should act in a multiphysics way, also considering—besides optimal energy management—NOx and CO_2 emissions [15] and, more generally, the reduction of pollutant emissions into the environment [16].

The aim of this paper is to address the topics of both drivability enhancement and the reduction of pollutant emissions; thus, the development of a vehicle model that is able to predict and highlight all these aspects is required—as carried out in [17], where Gear Shift Patterns were optimized to fulfil multiple constraints in terms of customer requirements such as drivability, NVH performance, emissions, and fuel consumption. These last two aspects were considered in [18] to evaluate fuel savings and NOx benefits when predictive control is applied to a mild hybrid truck, using dynamic programming with backtracking. Conversely, [19] proposed a novel real-time Energy Management Strategy (EMS) that was integrated with a model to use physical considerations to estimate energy consumption during gearshifts in Dual-Clutch Transmissions (DCT).

The development of a transmission model with a high level of accuracy is necessary to describe the dynamic behaviour of the whole system well, although its complexity can be reduced with appropriate simplifications—as carried out in [20] with a hybrid model of a vehicle driveline that was partially lumped and partially distributed to investigate vibrational phenomena inside the transmissions known as "shuffle" and "clonk". A methodology for the evaluation of clunking noises during gearshifts was also proposed in [21] using a nonlinear lumped parameter model of DCT, developed in an Amesim environment, while a Matlab/Simulink environment was adopted in [22] for the evaluation of NVH performance in the DCT transmission of a C-segment passenger car. Coming back to hybrid vehicles, mathematical models of both conventional and mild hybrid powertrain were developed in [23] through the integration of a conventional manual transmission-equipped powertrain and an electric motor as a secondary power source, in order to study the performance of partial power-on gear shifts through the implementation of torque hole-filling by the electric motor during gearshifts.

Besides the modelling of mechanical components, an appropriate tyre model can have relevant effects on simulated dynamics, as they have also a considerable influence on damping effects inside the powertrain. [24] demonstrated, using appropriate identification techniques aimed at describing tyre torsional dynamics in the frequency range 10–50 Hz, that damping is more accurately represented if modelled as hysteretic and not as viscous when the tyre load is low. Conversely, [25] proposed three different tyre models, e.g., modelled as a simple torsional spring, adopting a linear slip model, or assuming a fixed relaxation length; these last two models introduced nonlinearity, as an instantaneous slip ratio was considered. The relaxation length was estimated from the steady state tyre properties in [26] and included wheel load and slip dependencies. The adoption of this parameter is necessary for the description of interactions between tyres and roads, as all the other chassis components influence vehicle dynamic properties through tyre contact forces and torques.

A lumped LuGre tyre ground contact model was modelled in [27] and in [28], where the developed dynamic friction tyre model was able to accurately capture the transient behaviour of the friction force observed during transitions between braking and acceleration phases. Other researchers have instead used software for tyre dynamics to compute the forces acting on the tyre and hence on the steel wheel for other purposes, such as fatigue assessment [29].

The parallel mild hybrid architecture of a heavy commercial vehicle developed for this research study is similar to the one adopted in [30], but with the improvement of implementing three different control strategies for the direct control of vehicle acceleration, vehicle speed control, and torque control, with superposed active damping of drive train oscillations. The torque distribution between the two power sources is managed as a different task from the regulation task (oscillation damping), adopting a specific control algorithm that does not require the implementation of an optimisation problem. Moreover, as regards the linear model used for the design of the controller, an accurate tyre model with a fixed relaxation length is adopted—thus decoupling the kinematic behaviour of the driving rear tyres from the free rolling condition, as carried out in [25].

The developed model should also be able to reduce drivetrain vibrations; different works addressing this aim can be found in literature. In [31], a control for vibration reduction was developed for hybrid vehicles and experimentally validated, with two different controllers that reduced both the vibrations due to engine torque ripple and ones related to the driveline. Conversely, the root locus technique [32] has been adopted for anti-jerk control designs capable of preserving the driveline from oscillations. To improve NVH performance, a magnetic differential for vehicle drivetrains was developed in [33], while a controller for driving the engine—in order to actively damp driveline oscillations—was proposed in [34] to improve drivability and passenger comfort.

The present paper aims to propose an accurate nonlinear vehicle model and a model-based torque allocation control to achieve both improved vehicle drivability and pollutant emissions performance. The controller is designed in two steps with different aims: high-level logic was used in order to aim at providing a total powertrain torque able to satisfy the desired drivability performance in terms of longitudinal acceleration dynamics, through the integration of a direct feedforward term with a feedback contribution designed based on the Linear Quadratic Regulator (LQR) theory [35–37]; meanwhile, the low-level control allocation is designed to split the torque between the engine and the electric motor for emissions reduction purposes. The main contribution of the paper is to:

- propose a nonlinear dynamic model that considers the main vehicle nonlinearities, e.g., the elastic and damping behaviour of the torsional damper and the transient model of the tyres' dynamics—thus representing a reference model for vehicle performance assessment and torque controller validation;
- show the benefits introduced by the proposed controller in terms of dynamic performance, driveline oscillations, and NOx emissions;
- test the controller's robustness against the presence of unpredictable external inputs, e.g., a sudden road slope.

The paper is organised as follows: in Section 2 the nonlinear truck model—developed in a Matlab/Simulink environment—is shown and its characteristic dynamic equations are described, while in Section 3, the torque allocation control—based on the driver's intentions—is designed to reduce driveline oscillations and improve emissions performance. In Section 4, the controller performance is investigated to analyse the dynamic response of the powertrain and to evaluate the reduction of engine emissions. Finally, the conclusions of the work are discussed in Section 5.

2. Hybrid Vehicle Powertrain Layout

In this paper, a P0 mild parallel hybrid electric powertrain with a diesel engine was adopted. In this system architecture, better known as a belted alternator starter (BAS) [38,39], the internal combustion engine (ICE) and the electric motor (EM) are combined by means of a belt and work together to supply the necessary power to the vehicle. The EM is not able to propel the vehicle by itself; hence, it helps the ICE when extra power is required, behaving as a power booster [10]. With this configuration, an efficient engine start/stop functionality was ensured as well as the storage of energy in the battery during regenerative braking. Then, the power unit system was connected to the wheels through a gearbox and a differential.

The implementation of a parallel mild hybrid allowed us to benefit from the relevant advantages of the cooperation of EM and ICE, such as emissions reductions in case of slower transient behaviour in the ICE—better known as phlegmatization [40]—while the EM covered possible aggressive driver torque requests thanks to its high bandwidth—typical of electrical machines.

2.1. Matlab/Simulink Nonlinear Truck Model

The truck, shown in Figure 1, was modelled by means of a block-oriented approach [41,42], using a customized library of driveline components and adopting a user-friendly graphical interface—as shown in Figure 2. For each system component, it is necessary to define the dynamic laws by means of mechanical equilibrium principles, defining dynamic constants and constraints, and imposing the initial conditions. The input/output quantities of each component depend on the component topology. The Simulink model was grouped into subsystems, i.e., the engine, clutch damper, gearbox, differential, and wheels (2 driving and 2 driven), and a longitudinal truck vehicle model with longitudinal load transfer computation.

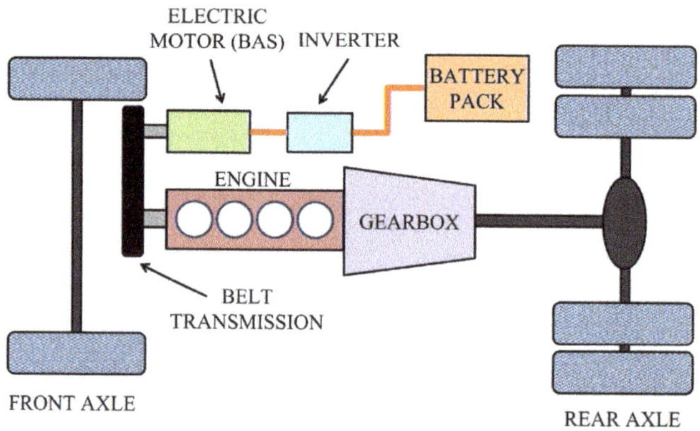

Figure 1. Transmission scheme of the parallel hybrid truck vehicle.

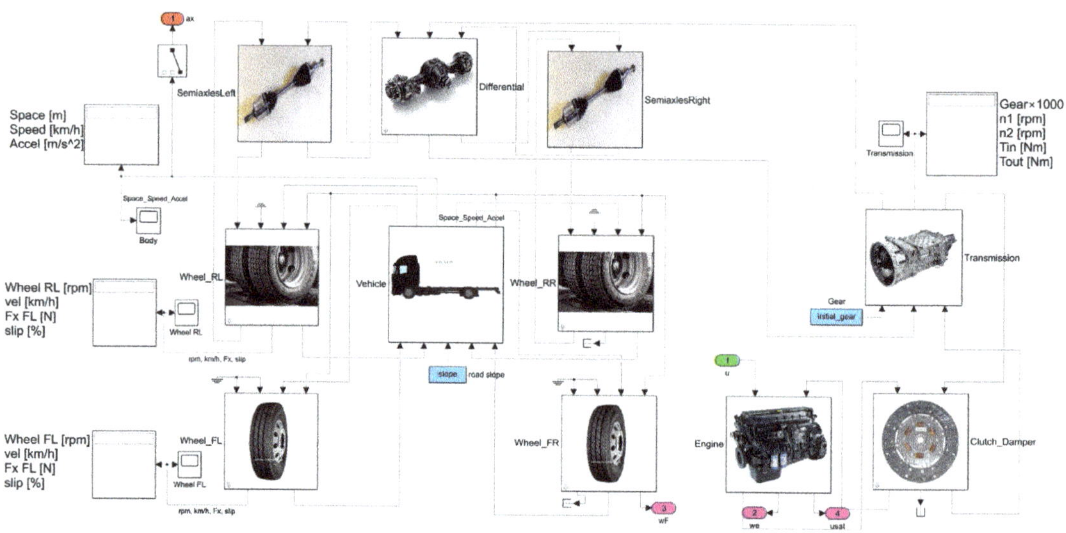

Figure 2. Nonlinear truck vehicle model in Matlab/Simulink environment.

Moreover, the model included the bi-linear elastic and damping behaviour of the torsional damper, the linear characteristic of the left and right-half shafts, a transient, nonlinear model for the tyres, and a Pacejka magic formulation [43] for road–tyre interactions.

Inside both the front and rear-tyre Simulink blocks, a wheel sensor was modelled, in order to apply a time delay and the effects of signal quantization to the simulated-wheel angular speed signal, to better reproduce experimental conditions.

2.2. Dynamic Equations

All the dynamic equations reported in this section are written in agreement with the free body diagrams of the drivetrain components and of the vehicle, which are reported in Figure 3.

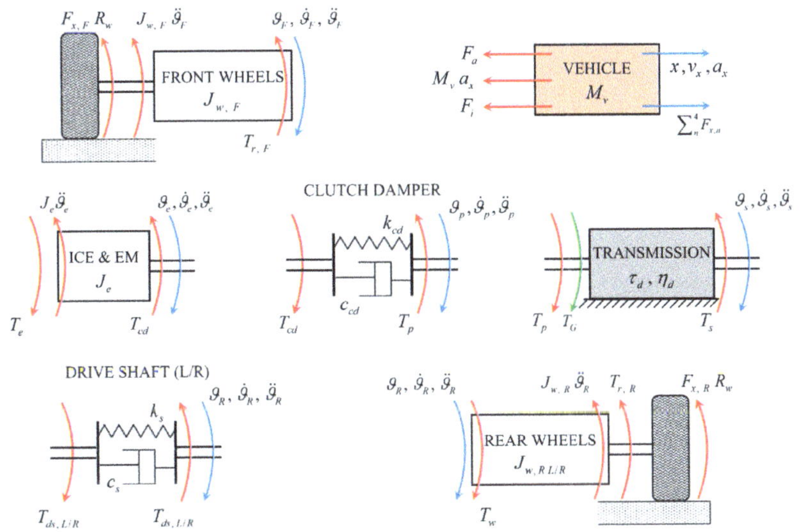

Figure 3. Free body diagrams of the main driveline components.

2.2.1. Power Source: ICE and EM

In the parallel hybrid configuration analysed, the ICE and the electric motor were connected by means of a belt, with a certain transmission ratio $\tau_b = \omega_{EM}/\omega_{ICE}$; hence, the total driving torque T_e, assuming 100% belt transmission efficiency, is given by:

$$T_e = T_{ICE} + T_{EM}\,\tau_b \tag{1}$$

The maximum torque of both the ICE and EM are reported in Figure 4. It is evident that the EM characteristic is symmetrical with respect to zero in case of traction or regenerative braking.

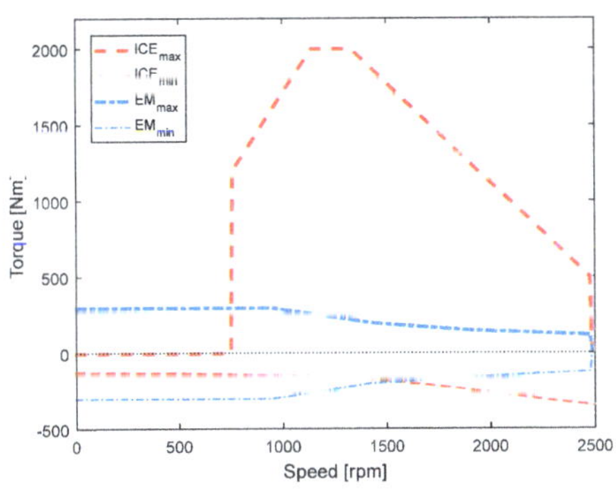

Figure 4. Torque characteristics of ICE and EM.

The acceleration of the engine $\ddot{\theta}_e$ was obtained from the equilibrium between the driving torque T_e, and the resistive torque of the primary shaft T_p, according to:

$$\ddot{\theta}_e = \frac{T_e - T_p}{J_e} \tag{2}$$

where J_e is the equivalent moment of inertia of the power source, so it depends on the ICE and EM inertias, which are related by the transmission ratio of the belted system as follows:

$$J_e = J_{ICE} + J_{EM}\,\tau_b^2 \tag{3}$$

2.2.2. Transmission: From Clutch Damper to Differential

The torque at the primary shaft T_p was obtained from the bi-linear clutch damper behaviour—whose nonlinear characteristics are reported in Figure 5—according to Equation (4):

$$T_p = T_{cd} = k_{cd}(\theta_e - \theta_p) + c_{cd}(\dot{\theta}_e - \dot{\theta}_p) \tag{4}$$

where k_{cd} and c_{cd} are the torsional stiffness and the torsional damping of the clutch damper, respectively.

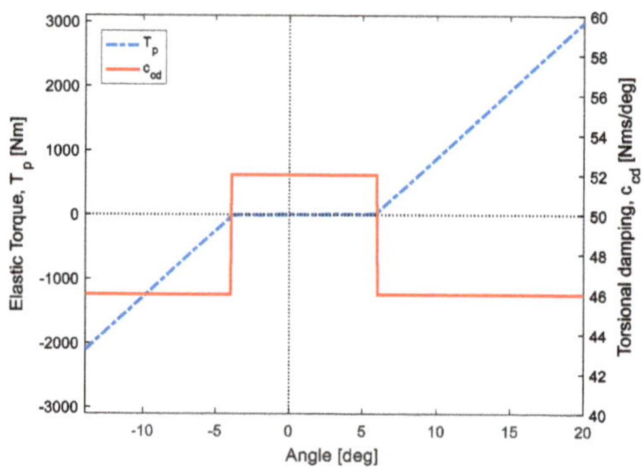

Figure 5. Elastic torque and viscous damping of the clutch damper.

The angular speed of the gearbox primary shaft $\dot{\theta}_p$ is related to the secondary shaft angular speed $\dot{\theta}_s$ by the kinematic relationship:

$$\dot{\theta}_p = \dot{\theta}_s\,\tau_d \tag{5}$$

where τ_d is the global driveline transmission ratio, depending on the actual gearbox gear ratio and the differential gear ratio, while the available torque at the secondary shaft—considering driveline efficiency η_d—is given by:

$$T_s = \eta_d\,\tau_d\,T_p = 2\,T_{ds,L/R} = 2[k_s(\theta_s - \theta_R) + c_s(\dot{\theta}_s - \dot{\theta}_R)] \tag{6}$$

As reported in Equation (6), the drive shaft torque $T_{ds,L/R}$ was lumped using a viscous damper c_s and a torsional spring k_s. Instead, the torque $T_G = T_s - T_p$ corresponds to the torque supported by the gearbox constraints. Finally, torque dissipations due to transmission bearings [44] were modelled, but not reported in the FBDs of Figure 3.

2.2.3. Vehicle and Wheels

The vehicle model considers the longitudinal dynamics and load transfer caused by the vehicle acceleration a_x, which is computed from the equilibrium equation—reported in Equation (7)—of the forces acting on the vehicle, i.e., the longitudinal forces on the tyres $F_{x,FL}$, $F_{x,FR}$, $F_{x,RL}$, $F_{x,RR}$, where the subscripts FL, FR, RL, RR refer to the front-left, front-right, rear-left and rear-right wheels, respectively, the aerodynamic force F_a and the resistance force F_i due to the road slope angle α.

$$a_x = \frac{\sum_n^4 F_{x,n} - F_a - F_i}{M_v} \tag{7}$$

The term $\sum_n^4 F_{x,n}$—where the subscript n refers to the *FL*, *FR*, *RL*, *RR* wheels—sums all the longitudinal forces on the four wheels, while the two terms F_a and F_i are as described by Equation (8):

$$\begin{cases} F_a = 0.5\,\rho\,S_v\,C_d\,V^2 \\ F_i = M_v\,g\,\sin\alpha \end{cases} \tag{8}$$

Regarding the aerodynamic terms, ρ is the air density, S_v is the vehicle front area, C_d is the aerodynamic drag coefficient, and V is the vehicle speed. Moreover, each tyre is described by an additional degree of freedom, and since vehicle lateral dynamics are not taken into account, the angular accelerations $\ddot{\theta}_F$ and $\ddot{\theta}_R$ are the same for the two wheels at the front and rear axles, and are computed according to Equation (9):

$$\begin{cases} \ddot{\theta}_F = \frac{-T_{b,F} - F_{x,F} R_w - T_{r,F}}{J_{w,F}} \\ \ddot{\theta}_R = \frac{T_w - T_{b,R} - F_{x,R} R_w - T_{r,R}}{J_{w,R}} \end{cases} \tag{9}$$

where $T_w = T_{ds,L/R}$ is the wheel torque coming from the half shaft—which participates only for the rear wheels—$T_{b,F}$ and $T_{b,R}$ are the braking torques, and $T_{r,n}$ ($T_{r,F}$ for the front and $T_{r,R}$ for the rear) is the rolling resistance torque, which has a quadratic relationship to the wheels' rotational speed θ_n, according to Equation (10):

$$T_{r,n} = F_{z,n}\,R_w\,(f + K\dot{\theta}_n^2) \tag{10}$$

The terms f and K are the constant and quadratic coefficients of the rolling resistance and $F_{z,n}$ is the vertical load on each wheel, which is computed considering the static load distribution between the two axles—by means of parameter γ—and the dynamic load transfer due to vehicle longitudinal acceleration, using Equation (11):

$$\begin{cases} F_{z,F} = \frac{m\,g\,\gamma}{2} - \frac{m\,h_G\,a_x}{2\,L} \\ F_{z,R} = \frac{m\,g\,(1-\gamma)}{2} + \frac{m\,h_G\,a_x}{2\,L} \end{cases} \tag{11}$$

where L is the vehicle wheelbase and h_G is the height of the vehicle's centre of mass, whereas $F_{z,F}$ and $F_{z,R}$ indicate the vertical load on the front and rear wheels, respectively. For each wheel, the longitudinal tyre slip is computed using Equation (12):

$$\upsilon = \frac{\omega R_w - V}{\omega R_w} \tag{12}$$

For sake of simplicity, the tyre model assumes a constant relaxation length L_t, even if—for more accurate results—it should be considered dependent on the vertical load and longitudinal slip [26]. Introducing the relaxation length, it is possible to correct the tyre slip, obtaining a transient slip for each tyre. Finally, the Pacejka formulation was adopted in the tyre model to evaluate the longitudinal force $F_{x,n}$ on each tyre, as a function of the different constant Pacejka parameters and three main dynamic quantities:

$$F_{x,n} = f(F_{z,n}, \mu, \sigma) \tag{13}$$

where μ is the tyre–road friction coefficient.

2.2.4. NOx Modelling

As proposed in [14], the NOx emissions were modelled considering the transient dynamics of the ICE. It has been experimentally proven that, in case of sudden acceleration by the driver, there is a spike in NOx emissions—which is expected to be 80% higher than the steady state value. Thus, in the Simulink model, a transfer function of the second order was adopted with a variable gain with respect to the engine torque T_{ICE}. The NOx emissions were estimated using the transfer function reported in Equation (14):

$$NO_X(s) = G(T_{ICE}(s))\ H(s)T_e(s) \tag{14}$$

where $G(T_{ICE}(s))$ is a variable gain between the steady state NOx and the engine torque, while the second order transfer function $H(s)$ is defined as:

$$H(s) = \frac{\omega_n^2}{s^2 + 2\zeta\omega_n + \omega_n^2} \tag{15}$$

where ω_n and ζ have been properly tuned to reproduce the transient dynamics relationship between the NOx emissions and the engine torque proposed by [14]. The NOx model here adopted takes into account only the engine-out emissions, without considering an after-treatment system device.

3. Closed Loop Control System

In this paper, the torque allocation control logic was designed to improve the vehicle drivability, to damp driveline oscillations, and to reduce engine emissions. The schematic diagram of the control strategy is reported in Figure 6. The total torque requested by the high-level part of the control logic T_e was obtained using the FeedForward (FF) contribution $T_{e,FF}$, supported by the FeedBack (FB) intervention $T_{e,FB}$. The total torque was then elaborated by the control allocation strategy (low level) to split the contribution between the engine T_{ICE} and the electric motor T_{EM} that were the input for the nonlinear model described in the previous section.

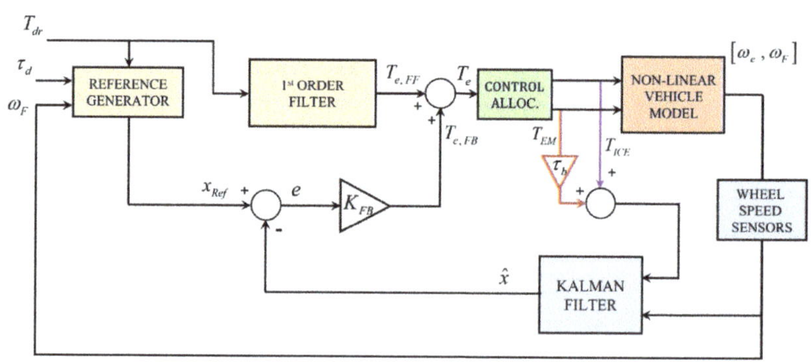

Figure 6. Driver control allocation control scheme.

3.1. *High-Level Control Strategy*

The gas pedal position (GPP) imposed by the driver was converted into a requested torque T_{dr} through the steady-state engine map in Figure 4 and then filtered out with a cut-off frequency of 2 Hz to elaborate the FF contribution $T_{e,FF}$.

The FB terms were designed based on a five-state linearised model, whose state space representation was defined according to the methodology presented in [11], where the effect of the torsional damper and load transfers was neglected with respect to the nonlinear model:

$$\{\dot{x}\} = [A]\{x\} + [B_1]v + [H] \tag{16}$$

where $v = T_e$ is the control output and $\{x\}$ is the state vector, composed as follows:

$$\{x\} = \begin{bmatrix} x_1 & x_2 & x_3 & x_4 & x_5 \end{bmatrix}^T = \begin{bmatrix} \frac{\theta_p}{\tau_d} - \theta_R & \dot{\theta}_R & \dot{\theta}_p & \dot{\theta}_F & T_t \end{bmatrix}^T \tag{17}$$

where $T_t = (F_{x,RL} + F_{x,RR})\ R_w$ is the steady-state driven wheel torque. Matrices $[A]$, $[B]$, and $[H]$ are described in [11]. The coordinate system of the state-space in Equation (16) is converted from the state vector $\{x\}$ of the error dynamics representation, as follows:

$$\{\dot{e}\} = [A]\{e\} + [B_1]\{v'\} \tag{18}$$

where $\{e\} = \{x_{Ref}\} - \{x\}$ and $v' = T_{e,FF} - v$, with $\{x_{Ref}\}$ represents the reference state vector, which is elaborated from the driver requested torque T_{dr}:

$$\{x_{Ref}\} = \begin{bmatrix} \frac{T_{dr}\tau_d}{k_s} & \dot{\theta}_{Ref} & \dot{\theta}_{Ref}\tau_d & \dot{\theta}_{Ref} & T_{dr}\tau_d \end{bmatrix}^T \tag{19}$$

The reference wheel speed $\dot{\theta}_{Ref}$ is obtained by integrating the reference acceleration $\ddot{\theta}_{Ref}$, defined based on the driver's intention (T_{dr} and τ_d), and the measured wheel speed $\dot{\theta}_{F,m}$:

$$\ddot{\theta}_{Ref} = \frac{T_{dr}\eta_d\tau_d - 0.5\ \rho\ S_v\ C_d\ (\dot{\theta}_{F,m}R_w)^2\ R_w - M_v g(f + K\ \dot{\theta}_{F,m}^2)R_w}{(J_{w,F} + J_{w,R} + M_v R_w^2 + J_e\eta_d\tau_d^2)} \tag{20}$$

The measured wheel speed $\dot{\theta}_{F,m}$, as mentioned in Section 2, was obtained using simulated sensor dynamics that introduced a communication time delay and the digitalization of the continuous measurement signal (zero order hold technique).

The FB torque contribution was then calculated as a full state feedback law:

$$v' = -[K_{FB}]\{e\} \tag{21}$$

where the feedback gains, saved in $[K_{FB}]$, are elaborated based on the LQR design—as proposed in [35,36]. The LQR control allowed the asymptotical stability of the closed loop system and the minimisation of the quadratic performance functional J:

$$J = \int_0^\infty (\{e\}^T[Q]\{e\} + rv'^2)dt \tag{22}$$

where $[Q] = diag(1e-9 \quad 1 \quad 1/\tau_d^2 \quad 1 \quad 1e-9)$ is a diagonal weighted matrix where the elements relative to angular speeds are emphasized with respect to the other states.

Finally, the feedback matrix $[K_{FB}]$ was obtained as:

$$[K_{FB}] = r^{-1}[B_1]^T[P] \tag{23}$$

where P is the solution of the algebraic Riccati equation.

The total requested torque was then formulated as:

$$T_e = v = T_{e,FF} - v' = T_{e,FF} + [K_{FB}]\{e\} \tag{24}$$

To implement a full state feedback controller, a Kalman filter was also designed to estimate the state vector $\hat{x}$, based on the inputs of the nonlinear vehicle model v and on its measured vector y.

$$\{y\} = \begin{bmatrix} \omega_e & \omega_F \end{bmatrix} = [C]\{x\} \tag{25}$$

where $[C] = \begin{bmatrix} 0 & 0 & 1 & 0 & 0; 0 & 0 & 0 & 1 & 0 \end{bmatrix}$ is the output matrix.

3.2. Control Allocation

The total torque T_e was then split between the ICE and the EM by the control allocation strategy. In this paper, the ICE provided the main torque contribution and the EM only intervened to compensate for the dynamic limits of the engine, e.g., when the driver requested a sudden torque through the GPP. The scheme of the control allocation strategy is reported in Figure 7.

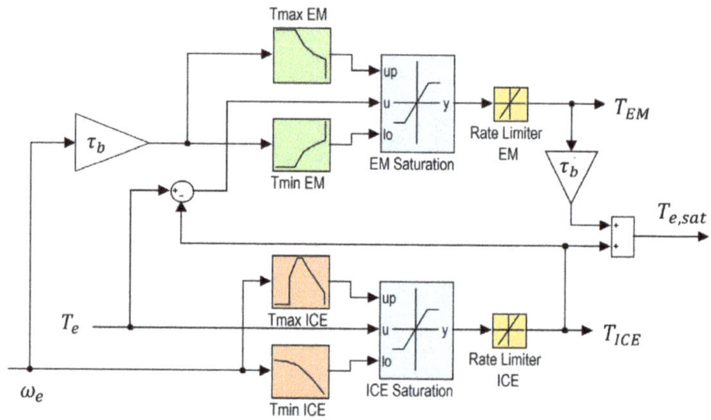

Figure 7. Control effort distribution between the ICE and EM, according to the dynamic limits of the engine.

As explained by several authors [14–16], the dynamic performance of the ICE is usually limited to reduce the engine emissions that may occur in presence of extremely dynamic torque requests. To accomplish the same task, a rate limiter was introduced into the control allocation strategy to saturate the maximum torque rate of the engine. Therefore, the torque of the EM, T_{EM}, is given by the difference between the total torque request T_e and the output torque requested to the ICE. As indicated in the scheme, the EM torque request is limited by a rate limiter, which was set at a higher level with respect to the ICE rate limiter.

In order to obtain the full benefits achieved by the methodology proposed in this paper, the energy stored in the battery should be sufficient to guarantee the requested activation of the electric motor. This energy management strategy is responsible for storing energy in the battery when the engine is more efficient, with lower polluting emissions. To exploit the benefits of this methodology, a more comprehensive design for the energy management system is required, but this is out of the scope of this paper.

4. Results

The torque allocation control designed in the previous section was finally applied to the nonlinear model described in Section 2 in order to evaluate its reference tracking performance, torque split efficacy between the ICE and the EM, and its influence on NOx emissions.

4.1. Reference Tracking Performance

The main purpose of the torque allocation control is to satisfy drivers' requests in terms of vehicle longitudinal acceleration. The driver's intention, expressed in terms of a reference longitudinal acceleration $a_{x,ref} = \ddot{\theta}_{Ref} R_w$, is converted into a total torque request T_e, which is obtained by the sum of the FF and FB contributions. The simulation results obtained during a tip-in manoeuvre—i.e., an instantaneous wheel torque request—are shown in Figures 8 and 9, where the influence of the two controller contributions is highlighted.

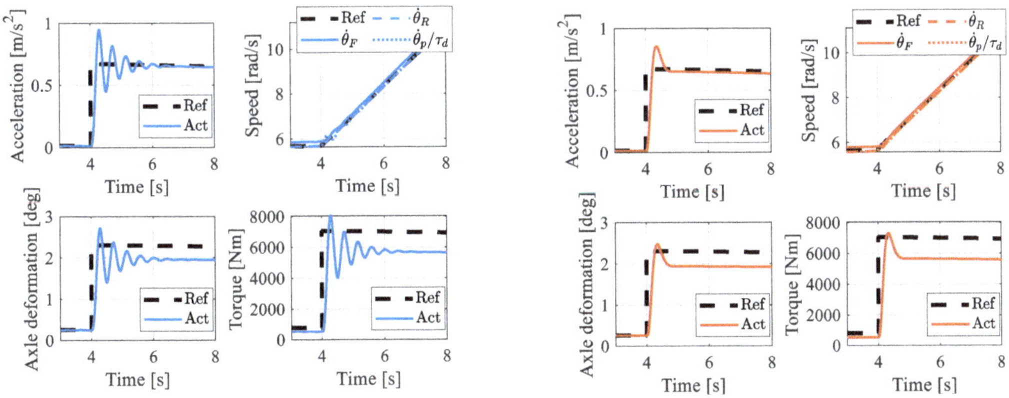

(**a**) (**b**)

Figure 8. Vehicle state tracking performance during a tip-in manoeuvre with only the FF torque contribution (**a**) and with FF + FB integration (**b**): longitudinal acceleration (top left), front wheels, rear wheels, and primary shaft speeds (top right), rear axle deformation (down left) and torque T_t.

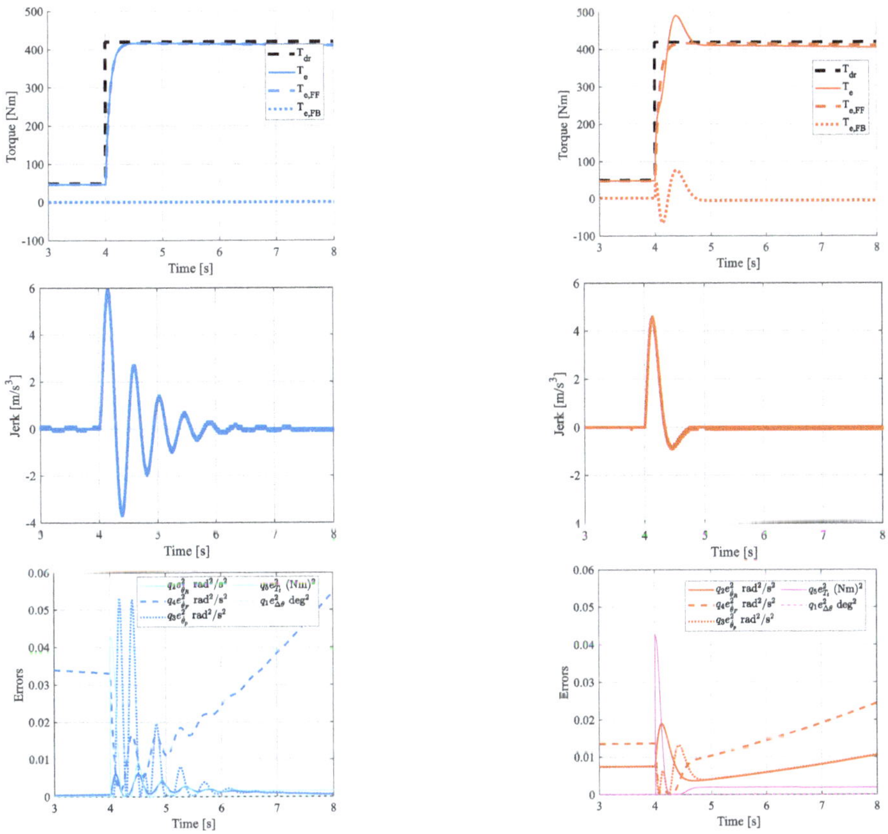

(**a**) (**b**)

Figure 9. Control torque request (top), jerk (centre) and weighted states errors (down) during a tip-in manoeuvre with only the FF torque contribution (**a**) and with FF + FB integration (**b**).

The control, with and without the activation of the FB contribution, provided a satisfactory reference tracking performance for the speed states. However, the presence of the FB contribution improved the transient vehicle response, with an evident oscillation reduction for the whole set of states. This also implies a considerable improvement in vehicle acceleration performance and in the consequent amplitude of the corresponding jerk (i.e., derivative of the vehicle acceleration), with beneficial implications for vehicle comfort. The advantages provided by the FB integration were paid off with a peak torque request of about 20% higher than that of controller with only the FF term activated.

The FB intervention also represented a fundamental contribution to the rejection of external disturbances that may affect controller tracking performance. Indeed, the FF formulation was based on the simplified linear model, and did not include motion resistance increments due to sudden road slopes. The influence of step disturbances on the road slope is analysed in Figure 10.

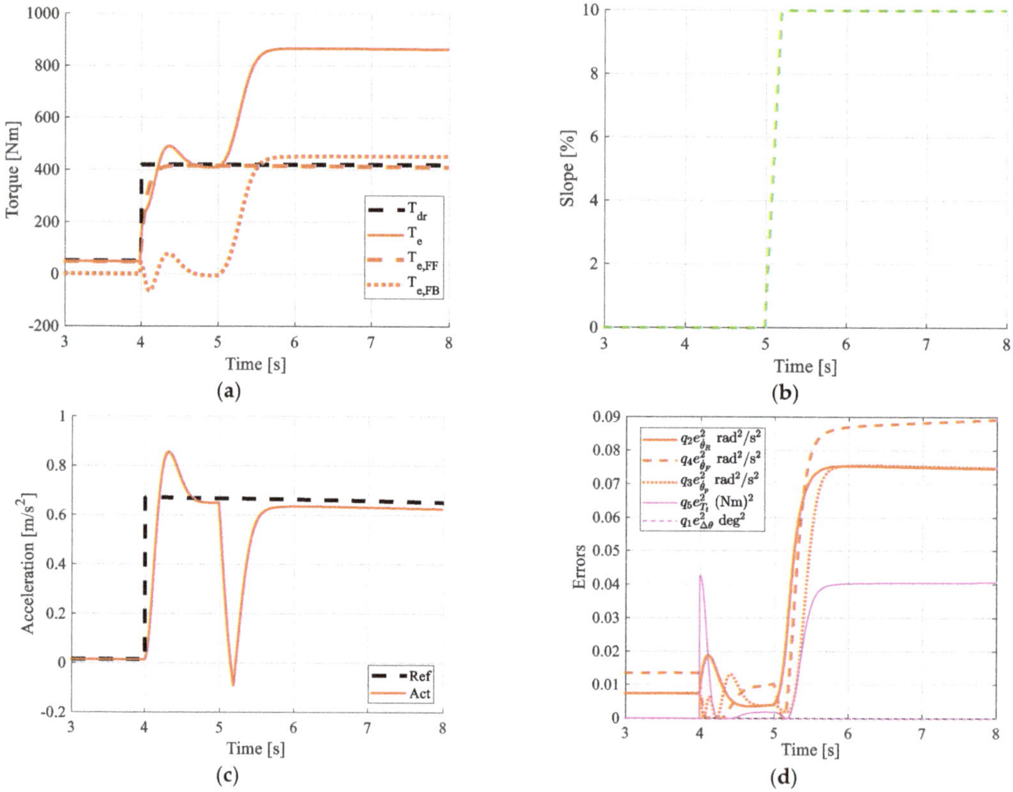

Figure 10. Control torque request (**a**), road slope (**b**), longitudinal acceleration (**c**) and weighted states errors (**d**) during a tip-in manoeuvre with the FF + FB controller mode.

The application of the torque requested by only the FF term would compromise the desired performance of the torque controller if external disturbances or uncertain/unmodelled dynamics affected the vehicle dynamics. Indeed, the FF formulation did not include the influence of the road slope, since this is a quantity that is difficult to measure or estimate—thus not representing a suitable input for feeding into a control logic. For this reason, the intervention of the FB term is essential to compensate for the higher vehicle motion resistance by applying an additional torque contribution that is able to bring the longitudinal acceleration back to the reference value; this was automatically achieved with an acceptable worsening of state errors and without any information related to the road profile.

The effect of the FB contribution was, however, influenced by the design of the LQR cost function and the selection of the weights q_i and r. The optimal control theory [35,36] suggests that the ratio q_i/r is responsible for the trade-off between reference tracking performance and energy consumption. A sensitivity analysis of the weight r on the torque controller performance is shown in Figure 11.

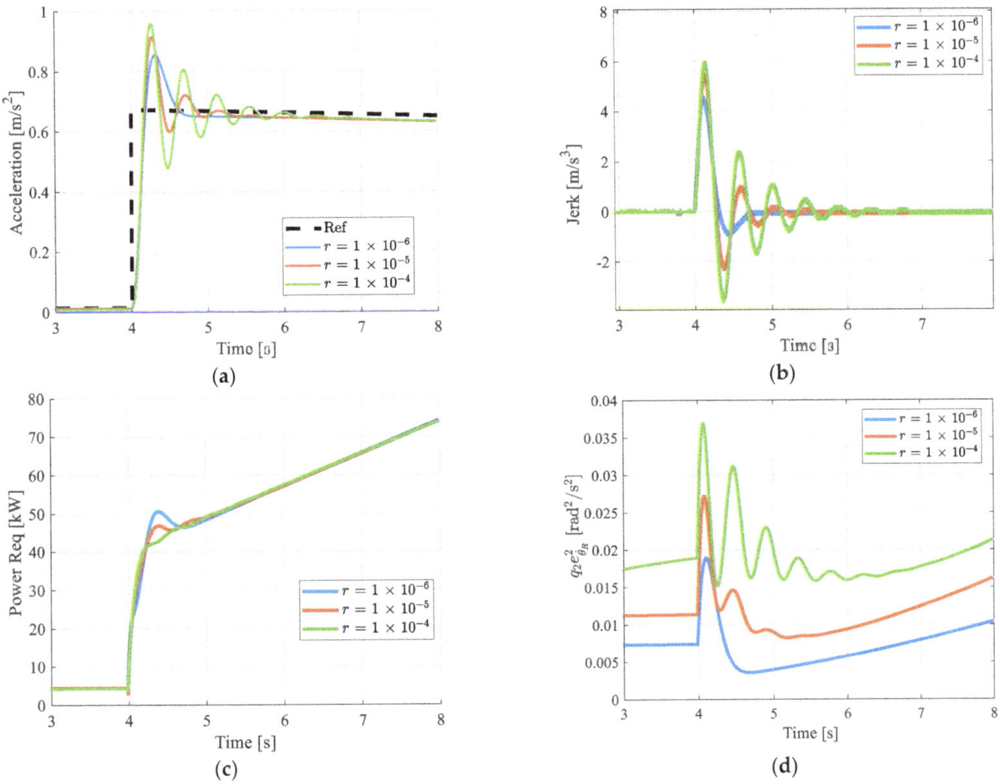

Figure 11. Vehicle acceleration (**a**), jerk (**b**), total power requested (c) and weighted rear speed error (**d**) during a tip-in manoeuvre with different *r* weights.

For low r values, the FB contribution aims at improving the reference tracking performance by reducing state errors and the amplitude of the vehicle acceleration and jerk oscillations. This advantage is obtained with a transient peak power request, i.e., the product $T_e\,\dot{\theta}_e$ of the engine and the electric motor. Vice versa, an increment of r would save more energy by reducing the total power request, but with an evident decline in terms of vehicle performance and comfort.

4.2. Control Allocation Evaluation

The previous section focused attention on the performance of the FF + FB controller in order to ascertain a reference total torque T_e that is able to dampen oscillations in vehicle acceleration, even in the presence of external disturbances, e.g., a road slope. However, the torque controller is not designed to take into account the powertrain's limits in terms of engine/electric motor torque saturation and dynamic constraints. Indeed, the ICE is usually operated at low frequency dynamics to reduce combustion emissions when a steep torque is requested, e.g., during a tip-in manoeuvre. The presence of a redundant driving source, i.e., the EM, allows the ICE's limitations to be overcome in terms of combustion emissions without compromising the vehicle's drivability. To evaluate the influence of ICE dynamics on the torque controller performance, a sensitivity analysis of the ICE rate limiter (see the scheme in Figure 7) is reported via the simulation results shown in Figures 12 and 13.

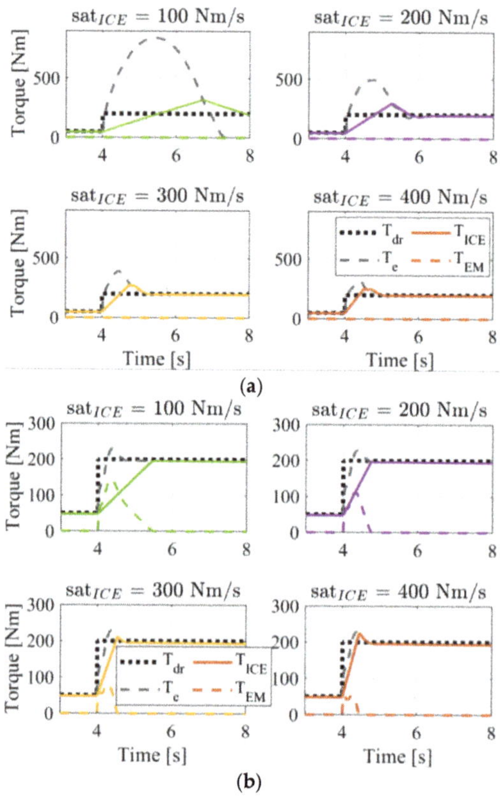

Figure 12. Torque split during a tip-in manoeuvre with only the ICE (**a**) and with a hybrid configuration (**b**) for different constraints on the ICE torque slope saturation.

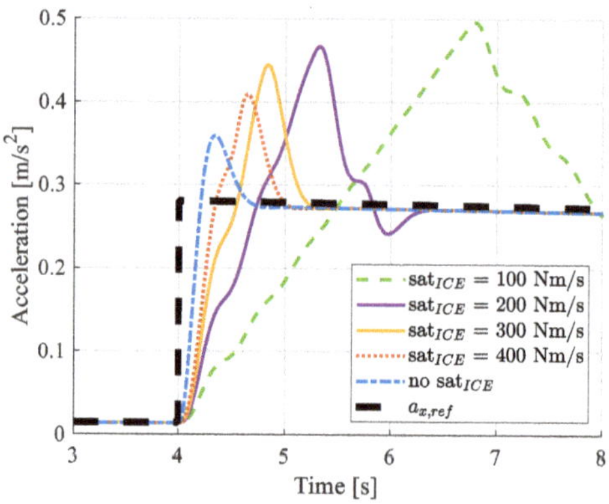

(**a**)

Figure 13. *Cont.*

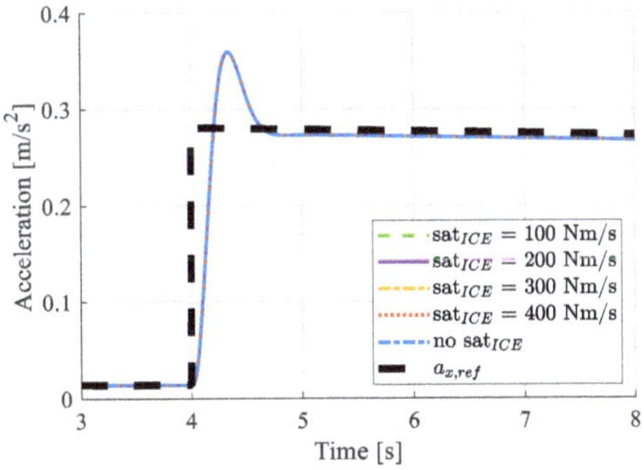

(b)

Figure 13. Longitudinal acceleration during a tip-in manoeuvre with only the ICE (**a**) and with a hybrid configuration (**b**) for different constraints on the ICE torque slope saturation.

The results presented in Figure 13a show how the dynamics of vehicle acceleration are affected by the rate limitation on the ICE-requested torque T_{ICE}. The lower the rate constraint on the T_{ICE}, the higher the overshoot and the rising time of the acceleration. These results are well explained, as the rate saturation on the torque requested by the ICE drastically influences the total torque T_e—see Figure 12a—demanded by the controller, which acts in order to reduce the states errors and so the acceleration error. The presence of an electric machine, with a typically faster response and lower emission issues than a combustion engine, allows the ICE to operate at lower frequency dynamics—thus achieving the desired performance in terms of vehicle acceleration, as shown in Figure 13b. The performance of the hybrid configuration is totally unaffected by limitations on the engine torque rate.

An important consequence of the results presented in Figures 12 and 13 is that the redundancy of power sources can also produce effective advantages in terms of the NOx emitted by the ICE. Indeed, assuming that the correlation between NOx emissions and the ICE output torque is expressed by Equation (15) [14], the benefits of reduced ICE dynamics are reported in Figure 14.

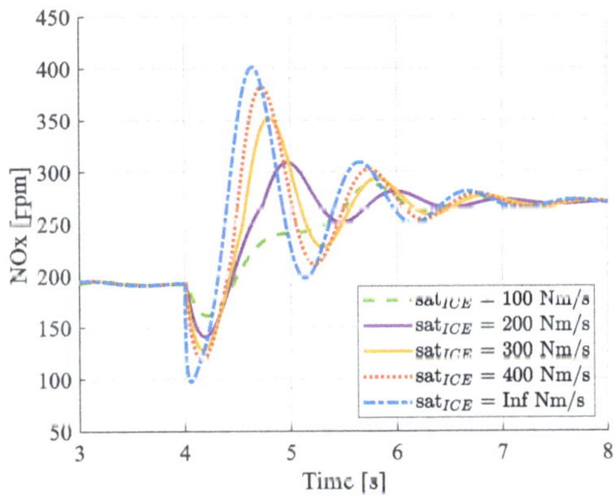

Figure 14. NOx emissions during a tip-in manoeuvre with the hybrid configuration for different constraints on the ICE torque slope saturation.

The hybrid configuration is then able to achieve the desired drivability performance, with a reduction of almost 80% in the transient NOx overshoot peak for the case of a 100 Nm/s torque rate saturation.

5. Conclusions

The paper presents a methodology for designing a torque allocation control strategy aiming at improving vehicle performance and NOx emissions. The results obtained can be summarized in the following final remarks:

- The high level of the torque control logic generated a total torque demand that satisfied the performance requirements in terms of vehicle speed and acceleration. Both the FF and FB contributions were designed based on a simplified version of the more accurate non-linear model. The FB contribution was fundamental in improving the vehicle's transient response, damping the acceleration and jerk oscillations. Differing calibrations of the FB gains can cope with different trade-offs between state error tracking performance and the power required to minimize errors;
- The paper also showed how the FB contribution was effective in rejecting external disturbances, e.g., the road slope, by compensating for the FF contribution with an additional contribution—thus satisfying the desired reference tracking performance;
- The control allocation strategy proved to produce a satisfactory vehicle drivability performance, even in the presence of tighter constraints in the ICE torque rate. The hybrid architecture showed outstanding robustness properties against variations in the ICE torque rate when compared to the ICE-only configuration. The redundancy offered by the fast dynamics of an electric machine represents an effective way of establishing the best combination between emissions and dynamic performance.

The results confirmed the methodology's efficacy for vehicle drivability performance and engine-out emissions. However, there is still potential scope for exploring the impact of the torque control strategy on other important features related to hybrid powertrains, e.g., energy management performance, clutch engagement, and regenerative braking benefits, which are reserved for future activities.

Author Contributions: Conceptualization, L.D., A.T., E.G. and M.V.; methodology, L.D., A.T., E.G. and M.V.; software, L.D., A.T., E.G. and M.V.; validation, L.D., A.T., E.G. and M.V.; formal analysis, L.D., A.T., E.G. and M.V.; investigation, L.D., A.T., E.G. and M.V.; writing—original draft preparation, L.D., A.T., E.G. and M.V.; writing—review and editing, L.D., A.T., E.G. and M.V.; visualization, L.D., A.T., E.G. and M.V.; supervision, E.G. and M.V. All authors have read and agreed to the published version of the manuscript.

Funding: This research received no external funding.

Institutional Review Board Statement: Not applicable.

Informed Consent Statement: Not applicable.

Data Availability Statement: Not applicable.

Conflicts of Interest: The authors declare no conflict of interest.

References

1. Awadallah, M.; Tawadros, P.; Walker, P.; Zhang, N. Comparative fuel economy, cost and emissions analysis of a novel mild hybrid and conventional vehicles. *Proc. IMechE Part D J. Automob. Eng.* **2018**, *232*, 1846–1862. [CrossRef]
2. Emadi, A.; Rajashekara, K.; Williamson, S.S.; Lukic, S.M. Topological overview of hybrid electric and fuel cell vehicular power system architectures and configurations. *IEEE Trans. Veh. Technol.* **2005**, *54*, 763–770. [CrossRef]
3. Sciarretta, A.; Back, M.; Guzzella, L. Optimal control of parallel hybrid electric vehicles. *IEEE Trans. Control Syst. Technol.* **2004**, *12*, 352–363. [CrossRef]
4. Barsali, S.; Miulli, C.; Possenti, A. A control strategy to minimize fuel consumption of series hybrid electric vehicles. *IEEE Trans. Energy Convers.* **2004**, *19*, 187–195. [CrossRef]
5. Castellano, A.; Cammalleri, M. Power losses minimization for optimal operating maps in power-split HEVs: A case study on the Chevrolet Volt. *Appl. Sci.* **2021**, *11*, 7779. [CrossRef]
6. Tota, A.; Galvagno, E.; Velardocchia, M. On the power-weighted efficiency of multimode powertrains: A case study on a two-mode hybrid system. In *I4SDG 2021, Proceedings of the 1st Workshop IFToMM for Sustainable Development Goals, Mechanisms and Machine Science, Online, 25–26 November 2021*; Quaglia, G., Gasparetto, A., Petuya, V., Carbone, G., Eds.; Springer: Cham, Switzerland, 2022; Volume 108, pp. 522–531. [CrossRef]

7. Tota, A.; Galvagno, E.; Dimauro, L.; Vigliani, A.; Velardocchia, M. Energy management strategy for hybrid multimode powertrains: Influence of inertial properties and road inclination. *Appl. Sci.* **2021**, *11*, 11752. [CrossRef]
8. Mantriota, G.; Reina, G.; Ugenti, A. Performance evaluation of a compound power-split CVT for hybrid powertrains. *Appl. Sci.* **2021**, *11*, 8749. [CrossRef]
9. Cirimele, V.; Dimauro, L.; Repetto, M.; Bonisoli, E. Multi-objective optimisation of a magnetic gear for powertrain applications. *Int. J. Appl. Electromagn. Mech.* **2019**, *60*, S25–S34. [CrossRef]
10. Nazari, S.; Siegel, J.; Stefanopoulou, A. Optimal energy management for a mild hybrid vehicle with electric and hybrid engine boosting systems. *IEEE Trans. Veh. Technol.* **2019**, *68*, 3386–3399. [CrossRef]
11. Galvagno, E.; Velardocchia, M.; Vigliani, A. Drivability enhancement and transient emission reduction for a mild hybrid diesel-electric truck. *Int. J. Powertrains* **2013**, *2*, 262–291. [CrossRef]
12. Parmar, V.; Di Rocco, D.; Sopouch, M.; Albertini, P. Multi-physics simulation model for noise and vibration effects in hybrid vehicle powertrain. *SAE Tech. Pap.* **2014**, 2014-01-2093. [CrossRef]
13. Chen, J.-S.; Hwang, H.-Y. Engine automatic start–stop dynamic analysis and vibration reduction for a two-mode hybrid vehicle. *Proc. IMechE Part D J. Automob. Eng.* **2013**, *227*, 1303–1312. [CrossRef]
14. Hagena, J.R.; Filipi, Z.S.; Assanis, D.N. Transient diesel emissions: Analysis of engine operation during a tip-in. *SAE Tech. Pap.* **2006**, 2006-01-1151. [CrossRef]
15. Wüst, M.; Krüger, M.; Naber, D.; Cross, L.; Greis, A.; Lachenmaier, S.; Stotz, I. Operating strategy for optimized CO_2 and NOx emissions of diesel-engine mild-hybrid vehicles. In Proceedings of the 15. Internationales Stuttgarter Symposium: Automobil- und Motorentechnik, Stuttgart, Germany, 17–18 March 2015; Springer: Wiesbaden, Germany, 2015. [CrossRef]
16. Thibault, L.; Sciarretta, A.; Degeilh, P. Reduction of pollutant emissions of diesel mild hybrid vehicles with an innovative energy management strategy. In Proceedings of the 2017 IEEE Intelligent Vehicles Symposium (IV), Los Angeles, CA, USA, 11–14 June 2017; pp. 1274–1279. [CrossRef]
17. Le Guen, D.; Weck, T.; Balihe, A.; Verbeke, B. Definition of gearshift pattern: Innovative optimization procedures using system simulation. *SAE Int. J. Engines* **2011**, *4*, 412–431. [CrossRef]
18. Pramanik, S.; Anwar, S. Predictive optimal control of mild hybrid trucks. *Vehicles* **2022**, *4*, 71. [CrossRef]
19. Guercioni, G.R.; Galvagno, E.; Tota, A.; Vigliani, A. Adaptive equivalent consumption minimization strategy with rule-based gear selection for the energy management of hybrid electric vehicles equipped with dual clutch transmissions. *IEEE Access* **2020**, *8*, 190017–190038. [CrossRef]
20. Farshidianfar, A.; Ebrahimi, M.; Bartlett, H. Hybrid modelling and simulation of the torsional vibration of vehicle driveline systems. *Proc. IMechE Part D J. Automob. Eng.* **2001**, *215*, 217–229. [CrossRef]
21. Galvagno, E.; Dimauro, L.; Mari, G.; Velardocchia, M.; Vella, A.D. Dual Clutch Transmission vibrations during gear shift: A simulation-based approach for clunking noise assessment. *SAE Tech. Pap.* **2019**, 2019-01-1553. [CrossRef]
22. Galvagno, E.; Guercioni, G.R.; Vigliani, A. Sensitivity analysis of the design parameters of a Dual-Clutch Transmission focused on NVH performance. *SAE Tech. Pap.* **2016**, 2016-01-1127. [CrossRef]
23. Awadallah, M.; Tawadros, P.; Walker, P.; Zhang, N. Dynamic modelling and simulation of a manual transmission based mild hybrid vehicle. *Mech. Mach. Theory* **2017**, *112*, 218–239. [CrossRef]
24. Guzzomi, A.L.; Sharman, A.; Stone, B.J. Some torsional stiffness and damping characteristics of a small pneumatic tyre and the implications for powertrain dynamics. *Proc. IMechE Part D J. Automob. Eng.* **2010**, *224*, 229–244. [CrossRef]
25. Bartram, M.; Mavros, G.; Biggs, S. A study on the effect of road friction on driveline vibrations. *Proc. IMechE Part K J. Multi-Body Dyn.* **2010**, *224*, 321–340. [CrossRef]
26. Rill, G. First order tyre dynamics. In Proceedings of the III European Conference on Computational Mechanics, Solids, Structures and Coupled Problems in Enginnering, Lisbon, Portugal, 5–8 June 2006. [CrossRef]
27. Dolcini, P.J.; Canudas-de-Wit, C.; Béchart, H. *Dry Clutch Control for Automotive Applications*, 1st ed.; Springer: London, UK, 2010; pp. 1–144.
28. Canudas-de-Wit, C.; Tsiotras, P.; Velenis, E.; Basset, M.; Gissinger, G. Dynamic Friction Models for Road/tyre Longitudinal Interaction. *Veh. Syst. Dyn.* **2003**, *39*, 189–226. [CrossRef]
29. Rovarino, D.; Actis Comino, L.; Bonisoli, E.; Rosso, C.; Venturini, S.; Velardocchia, M.; Baecker, M.; Gallrein, A. Hardware and virtual test-rigs for automotive steel wheels design. *SAE Int. J. Adv. Curr. Prac. Mobil.* **2020**, *2*, 3481–3489. [CrossRef]
30. Friedriksson, J. Improved driveability of a hybrid electric vehicle using powertrain control. *Int. J. Altern. Propuls.* **2006**, *1*, 97–111. [CrossRef]
31. Ito, Y.; Tomura, S.; Moriya, K. Development of vibration reduction motor control for hybrid vehicles. In Proceedings of the IECON 2007 33rd Annual Conference of the IEEE Industrial Electronics Society, Taipei, Taiwan, 5–8 November 2007; pp. 516–521. [CrossRef]
32. Baumann, J.; Torkzadeh, D.D.; Ramstein, A.; Kiencke, U.; Schlegl, T. Model-based predictive anti-jerk control. *Control Eng. Pract.* **2006**, *14*, 259–266. [CrossRef]
33. Filippini, M.; Torchio, R.; Alotto, P.; Bonisoli, E.; Dimauro, L.; Repetto, M. A new class of devices: Magnetic gear differentials for vehicle drivetrains. *IEEE Trans. Transp. Electrif.* **2022**. [CrossRef]
34. Berriri, M.; Chevrel, P.; Lefebvre, D. Active damping of automotive powertrain oscillations by a partial torque compensator. *Control Eng. Pract.* **2008**, *16*, 874–883. [CrossRef]

35. Anderson, B.D.O.; Moore, J.B. *Optimal Control: Linear Quadratic Methods*, 1st ed.; Prentice-Hall: Englewood Cliffs, NJ, USA, 1990.
36. Ostertag, E. *Mono- and Multivariable Control and Estimation: Linear, Quadratic and LMI Methods*; Springer: Berlin/Heidelberg, Germany, 2011.
37. Venturini, S.; Bonisoli, E. Design of a spherical pendulum didactic test rig. *Int. J. Mech. Control* **2018**, *19*, 69–76.
38. Walters, J.E.; Krefta, R.J.; Gallegos-Lopes, G.; Fattic, G.T. Technology considerations for belt alternator starter systems. *SAE Tech. Pap.* **2004**, 2004-01-0566. [CrossRef]
39. Morra, E.; Spessa, E.; Ciaravino, C.; Vassallo, A. Analysis of various operating strategies for a parallel-hybrid diesel powertrain with a belt alternator starter. *SAE Int. J. Altern. Powertrains* **2012**, *1*, 231–239. [CrossRef]
40. Auerbach, M.; Ruf, M.; Bargende, M.; Reuss, H.-C.; Van Doorn, R.; Wilhelm, F.; Kutschera, I. Potentials of phlegmatization in diesel hybrid electric vehicles. *SAE Tech. Pap.* **2011**, 2011-37-0018. [CrossRef]
41. Velardocchia, M.; D'Alfio, N.; Bonisoli, E.; Galvagno, E.; Amisano, F.; Sorniotti, A. Block-oriented models of torque gap filler devices for AMT transmissions. *SAE Tech. Pap.* **2008**, 2008-01-0631. [CrossRef]
42. Dimauro, L.; Bonisoli, E.; Repetto, M. Dynamic behaviour and magneto-mechanical efficiency of a contactless magnetic transmission. In *Rotating Machinery, Optical Methods & Scanning LDV Methods, Proceedings of the 40th IMAC, Orlando, FL, USA, 7–10 February 2022*; Di Maio, D., Baqersad, J., Eds.; Springer: Cham, Switzerland, 2023; Volume 6, pp. 129–138. [CrossRef]
43. Pacejka, H.B. *Tyre and Vehicle Dynamics*, 2nd ed.; Butterworth: Oxford, UK, 2005; pp. 1–621.
44. Sorniotti, A.; Sampò, E.; Velardocchia, M.; Bonisoli, E.; Galvagno, E. Friction inside wheel hub bearings: Evaluation through analytical models and experimental methodologies. *SAE 2007 Trans. J. Engines* **2007**, *116*, 1665–1676. [CrossRef]

applied sciences

MDPI

Article

Battery High Temperature Sensitive Optimization-Based Calibration of Energy and Thermal Management for a Parallel-through-the-Road Plug-in Hybrid Electric Vehicle

Pier Giuseppe Anselma [1,2,*], Marco Del Prete [1] and Giovanni Belingardi [1,2]

[1] Department of Mechanical and Aerospace Engineering (DIMEAS), Politecnico di Torino, 10129 Torino, Italy; delpretemarco96@gmail.com (M.D.P.); giovanni.belingardi@polito.it (G.B.)
[2] Center for Automotive Research and Sustainable Mobility (CARS), Politecnico di Torino, 10129 Torino, Italy
* Correspondence: pier.anselma@polito.it

Abstract: Preserving high-voltage battery pack lifetime represents a key issue in hybrid electric vehicles (HEVs). Temperature has remarkably major impacts on battery lifetime and implementing HEV thermal and energy management approaches to enhance fuel economy while preserving battery lifetime at various temperatures still represents an open challenge. This paper introduces an optimization driven methodology to tune the parameters of thermal and energy on-board rule-based control approaches of a parallel through-the-road plug-in HEV. Particle swarm optimization is implemented to this end and the calibration objective involves minimizing HEV operative costs concerning energy consumption and battery degradation over the entire vehicle lifetime for various ambient temperatures, driving conditions, payload conditions, and cabin conditioning system states. Numerical models are implemented that can estimate the evolution over time of the state of charge, state of health, and temperature of HEV high-voltage battery packs. Obtained results suggest that the calibrated thermal and energy management strategy tends to reduce pure electric operation as the ambient temperature progressively increases beyond 30 °C. The consequent longer internal combustion engine operation entails a gradual increase in the overall vehicle energy demand. At a 36 °C ambient temperature, the HEV consumes 2.3 times more energy compared with the 15 °C reference value. Moreover, activating the cabin conditioning system seems beneficial for overall plug-in HEV energy consumption at high ambient temperatures. The presented methodology can contribute to easing and accelerating the development process for energy and thermal management systems of HEVs.

Keywords: battery ageing; battery lifetime; battery temperature; calibration optimization; control strategy; electrified powertrain; energy management; hybrid electric vehicle; thermal management

Citation: Anselma, P.G.; Del Prete, M.; Belingardi, G. Battery High Temperature Sensitive Optimization-Based Calibration of Energy and Thermal Management for a Parallel-through-the-Road Plug-in Hybrid Electric Vehicle. *Appl. Sci.* **2021**, *11*, 8593. https://doi.org/10.3390/app11188593

Academic Editors: Marco Cammalleri and Vincenzo Di Dio

Received: 30 June 2021
Accepted: 11 September 2021
Published: 16 September 2021

1. Introduction

Lithium-ion batteries currently represent the most commonly employed energy storage systems on-board hybrid electric vehicles (HEVs) and pure electric vehicles [1,2]. They allow the storage of electrical energy for subsequent use as vehicle propelling energy. However, they present some critical issues that currently contribute to restricting the widespread diffusion of electrified powertrains. As an example, the current high cost of Li-ion battery packs (250–2500 $/kWh [3]) is a major factor in making the retail price of electrified vehicles considerably higher compared with vehicles equipped with traditional propulsion systems, i.e., based solely on internal combustion engines (ICEs) [4]. Moreover, Li-ion battery packs notably suffer from degradation of performance throughout their lifetime: their internal resistance increases, their capacity reduces, and their charge and discharge power capabilities consequently decrease [5]. Battery degradation worsens the performance of the electrified vehicle both in terms of fuel economy and acceleration capability [6]. Cyclic ageing that occurs during effective battery operation is normally considered the most critical ageing aspect [7]. The rate of cyclic ageing depends on given battery operating factors such as the C-rate, state-of-charge (*SOC*), and battery temperature as examples [8–10].

In general, the more critical the battery operating conditions, the more rapidly the battery approaches its end-of-life (*EOL*). However, defining the battery *EOL* is not a trivial operation. Indeed, the battery *EOL* is not simply related to the effective maximum number of charging and discharging cycles that a battery can supply, but rather to a more subtle concept of battery charge throughput (often expressed as Ah-throughput). The total number of allowed battery cycles loses relevance as the nominal capacity decreases below a certain value. The progression of the battery towards its *EOL* is normally monitored by the battery state-of-health (*SOH*) and it ends when the effective battery capacity is reduced to 80% of its initial nominal value [11].

Progressive capacity loss in Li-ion energy storage systems involves specific chemical processes that take place within the cells. For example, a solid electrolyte interface (SEI) layer gradually grows on the anode with a rate that particularly accelerates at high temperatures above 35 °C [12]. Battery ageing is also due to lithium plating at the anode, which principally occurs at low temperatures below 15 °C [13]. However, battery ageing also occurs when the temperature of the battery is between 15 °C and 35 °C at a rate that depends on the battery C-rate, *SOC* swing, and battery temperature itself [14].

Developing battery *SOH* sensitive EMSs for HEVs has become a topic of growing interest in the literature. In 2012, Ebbesen et al. first adopted the concept of *SOH* for controlling battery ageing using a throughput-based capacity fading model. They simulated different driving missions by means of a quasi-static HEV model controlled by an equivalent consumption minimization strategy (ECMS) [15]. Similar considerations were made by Onori et al. in the same year, who introduced a severity factor in place of the battery *SOH* [16]. Further battery *SOH* sensitive energy management approaches were then introduced in the following years. The concept of dual-objective ECMS for controlling fuel consumption and battery ageing was further extended in 2015 [17] and in 2019 [18]. These investigations demonstrated that considerable economic saving can be achieved by preserving the same battery pack for the entire vehicle lifetime. Different battery *SOH* sensitive HEV energy management approaches were developed, including convex optimization [19], model-predictive control [20], and dynamic programming [21,22].

The above studies assumed that battery temperature was kept constant by the conditioning system throughout the entire simulated driving missions. However, guaranteeing a constant battery temperature over time can represent a difficult task for the thermal regulation system, and slight temperature oscillations can occur even in best case scenarios. A number of studies have currently begun to assess this issue. The authors of this paper conducted a one-year long experimental campaign to validate the battery lifetime predicted by a *SOH* sensitive HEV EMS and found that an increase in the average cell surface temperature of only 1 °C can culminate in as much as a 15,000 km or 7% decrease in the cell lifetime [23]. Du et al. recently considered energy management for a parallel plug-in hybrid electric bus sensitive to both battery temperature and *SOH*. They suggested that noticeable equivalent cost savings might be attained by limiting the overall battery temperature increase while driving [24]. Although these works represent steps in the right direction, extensive research is still needed to develop a battery thermal management system whose calibration can guarantee a satisfactory high-voltage battery kilometrical lifetime. The present work introduces an optimization-driven calibration method for the on-board energy and thermal management of a plug-in HEV. The aim is to minimize operative costs of fuel and electrical energy consumption and battery degradation over the entire vehicle lifetime. The optimization-driven calibration workflow considers different ambient temperatures, driving conditions, payload conditions, and cabin conditioning system states. The developed HEV energy and thermal management strategy proved to flexibly adapt to different ambient temperatures. Moreover, the strategy limits the increase in the overall HEV energy demand that may be expected when ambient temperature rises above 30 °C.

The paper is organized as follows: the retained HEV powertrain model and baseline on-board EMS are presented first. Section 3 then describes the numerical models implemented to reproduce the electrical, thermal, and ageing behavior of the high-voltage HEV battery. The battery cooling system and its numerical model are also described. The optimization-driven calibration of the HEV EMS is discussed in Section 4. Results are presented in Section 5, with conclusions drawn in Section 6.

2. HEV Modeling and Baseline Energy Management

Figure 1 illustrates a schematic diagram of the parallel-through-the-road plug-in HEV powertrain retained in this study. The vehicle configuration has been inspired by the commercially available Jeep Renegade 4xe® [25]. This HEV is equipped with a 1.3-L, four-cylinder, turbocharged spark-ignition internal combustion engine (*ICE*) on the front axle. Moreover, two permanent magnet electric motor/generators are embedded, one in configuration P4 (at the rear axle) and one in configuration P0 as a belt starter-generator. These are respectively indicated as *MGP4* and *MGP0* in the paper. It should be noted that the high-voltage battery has been considered made of A123 26650 cells instead of $LiNiMnCoO_2$ (NMC) cells, as in the commercially available Renegade 4xe®. This choice stems from the availability of both a widely accepted battery cyclic ageing model and open-source data for A123 26650 cells [14,26]. The number of cells has been selected targeting overall pack parameters as close as possible to the ones of the commercially available plug-in HEV (e.g., 11.4 kWh nominal capacity and 400 V nominal voltage). Retained vehicle data are reported in Table 1, while the follow-up of this section discusses further details concerning the on-board HEV energy management logic and the considered powertrain modeling approach.

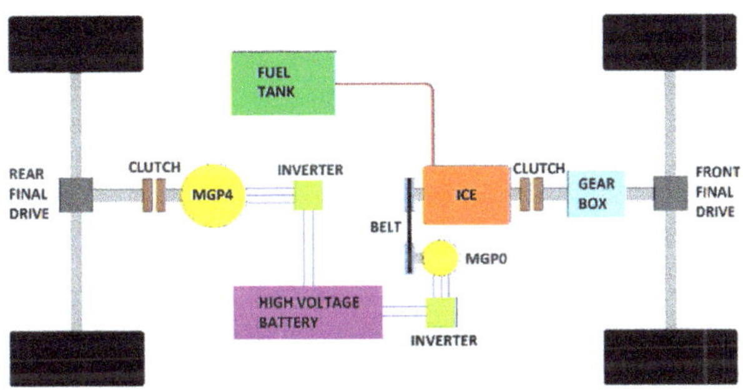

Figure 1. Schematic diagram of a parallel-through-the-road HEV powertrain.

Table 1. Parameters of the parallel-through-the-road plug-in HEV powertrain.

Component	Parameter	Value
ICE	Max. power (hp @ RPM)	130 @ 5500
	Max. torque (Nm @ RPM)	270 @ 1850
MGP4	Max. power (hp @ RPM)	60 @ >1240
	Max. torque (Nm @ RPM)	250 @ <1240
MGP0	Max. power (hp @ RPM)	20 @ >2480
	Max. torque (Nm @ RPM)	48 @ <2480
Transmission ratios	Rear final drive (-)	5.6
	Front final drive (-)	3.68
	Automatic gearbox (-)	[4.46; 2.51; 1.56; 1.14; 0.85; 0.67]
	MGP0 belt (-)	2.7

Table 1. *Cont.*

Component	Parameter	Value
High-voltage battery	Battery nominal capacity/voltage (kWh/V)	10.94/400
	Cell nominal capacity/voltage (Wh/V)	7.6/3.33
	Pack configuration	120 S; 12 P
	Cell type	A123 ANR26650
Vehicle body	Road load coefficient A (N)	94.04
	Road load coefficient B (Ns/m)	3.81
	Road load coefficient C (Ns^2/m^2)	0.48
	Wheel radius (m)	0.322

2.1. HEV on-Board Energy Management Strategy

The considered HEV embeds an *ICE*, a *MGP0*, and a *MGP4*. These three power components allow the HEV to choose between the three following operating modes while driving [25]:

1. Electric mode, in which the traction is provided by the *MGP4* alone;
2. Hybrid mode, which is characterized by the three power components being controlled in order to have the *ICE* working close to its optimal operating line (OOL). The OOL is particularly defined by the values of *ICE* torque that maximize the engine efficiency for each given value of rotational speed;
3. E-save mode, where the *ICE* can be simultaneously employed for propelling the HEV and charging the battery by means of *MGP0* and *MGP4* working as generators. In particular, the instantaneous *ICE* power is controlled to be the maximum value between the one corresponding to its OOL and the one required to propel the HEV alone.

All the operating modes allow performing regenerative braking, while the baseline HEV controller implemented in this study is assumed to operate as follows:

- As the HEV starts a given driving mission, pure electric operation is selected if the battery is sufficiently charged. Then, an automatic shift to Hybrid mode is performed if the *SOC* goes below 0.30 in order to preserve battery charge sustenance. Moreover, if the driver's power demand exceeds the value that can be provided by the *MGP4* alone in Electric mode, then the powertrain temporarily operates in Hybrid mode.
- In case the value of battery *SOC* falls below 0.25, the rule-based HEV controller switches to the E-save mode to charge the battery until *SOC* reaches 0.30. Hybrid mode is subsequently set as the operating mode.
- Regenerative braking is disabled when the battery *SOC* exceeds the upper limit, which is set to 0.8 here. This is performed as Li-Ion cells significantly degrades when being charged at high *SOC* values [27].

2.2. HEV Modeling Approach

A quasi-static approach was adopted for modeling the HEV in Matlab/Simulink® software (Mathworks Inc., Natick, MA, US). This approach is largely used in design and analysis activities for HEV powertrains thanks to its high computational efficiency [28]. A proportional–integral–derivative (PID) controller is introduced here to model the driver following the speed profile over time for given drive cycles. Lookup tables of power components employed in this paper are illustrated in Figure 2 and they have been generated using dedicated tools in Simcenter Amesim® software (Siemens PLM, Camberley, UK). The mathematical procedures implemented in the software to generate the lookup tables are available in the literature [29,30].

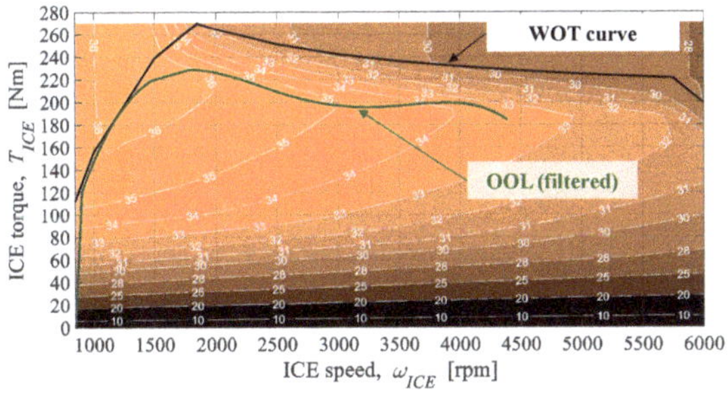

(a) ICE fuel efficiency map [%]

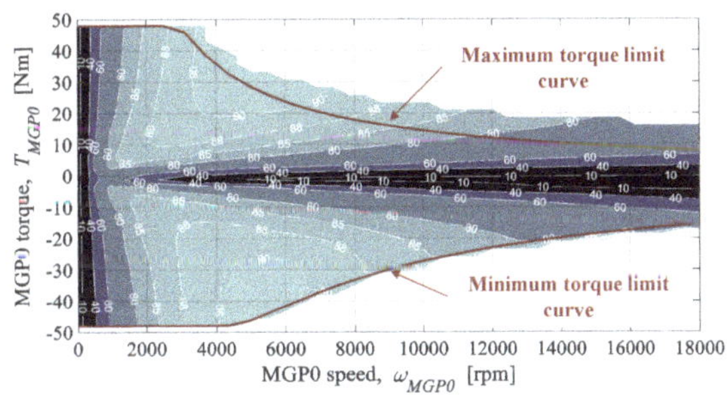

(b) MGP0 efficiency map [%]

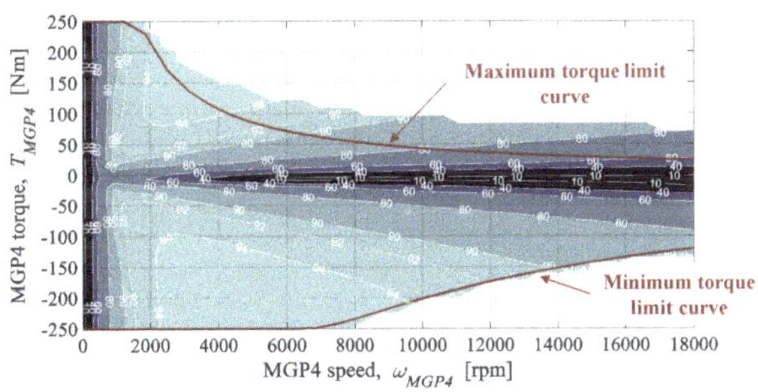

(c) MGP4 efficiency map [%]

Figure 2. Efficiency maps of *ICE*, *MGP0*, and *MGP4* generated by means of Simcenter Amesim® software.

The total vehicle resistive force F_{res} can be easily computed from the vehicle free body diagram using Equation (1):

$$F_{res} = F_{aero} + F_{roll} + F_{misc} + F_{grad} \tag{1}$$

where F_{aero} is the aerodynamic drag, F_{roll} represents the rolling resistance, F_{misc} incorporates some miscellaneous terms, whereas F_{grad} is the grading resistance related to road slope angle ϑ. Specifically, the above resistive terms were evaluated as follows:

$$F_{aero} + F_{roll} + F_{misc} = A + Bv + Cv^2 \tag{2}$$

$$F_{grad} = m_{veh} g sin\vartheta \tag{3}$$

where A, B, and C are the vehicle road load coefficients [31], v is the longitudinal speed of the vehicle, g represents the gravitational acceleration, and m_{veh} is the total mass of the vehicle in kilograms that was assumed to be related to the number of passengers n_{pass}:

$$m_{veh} = 1768 + 100 \cdot n_{pass}. \tag{4}$$

The tractive or braking force F_{pwt} that is provided at the wheels by the powertrain was introduced. Then, the equilibrium equation at the vehicle level could be written as follows:

$$m_{veh} a = F_{pwt} - F_{res,tot} \tag{5}$$

where a indicates the longitudinal acceleration of the vehicle. The powertrain torque T_{pwt} requested at the wheels could thus be computed as follows:

$$T_{pwt} = F_{pwt} r_{wheel} \tag{6}$$

where r_{wheel} is the radius of the wheels. All possible terms contributing to T_{pwt} were considered, and Equation (7) could thus be obtained:

$$T_{pwt} = T_{ICE,wheel} + T_{MGP4,wheel} + T_{MGP0,wheel} - T_{brakes} \tag{7}$$

where the terms $T_{ICE,wheel}$, $T_{MGP4,wheel}$, and $T_{MGP0,wheel}$ represent the values of torque provided at the wheels by the respective power components, while T_{brakes} is the braking torque supplied by the brake system.

The torques of power components T_{ICE}, T_{MGP4}, and T_{MGP0} were evaluated backwardly from the respective torques outputted at the wheels by means of relations using the transmission ratios reported in Table 1 and appropriate efficiencies [32]. The wheel angular speed ω_{wheel} could be defined as follows:

$$\omega_{wheel} = \frac{v}{r_{wheel}}. \tag{8}$$

ω_{MGP4}, ω_{ICE}, and ω_{MGP0} that respectively represent angular speeds of *MGP4*, *ICE*, and *MGP0* can be evaluated by multiplying times transmission ratios. Once the operating conditions (i.e., torque and speed) are known for each power component, evaluating the fuel consumed by the *ICE* and the electric power requested to the high-voltage battery by the electric motors is possible through considering the following equations:

$$M_{fuel} = \int_0^t \dot{m}_{fuel} dt + n_{crank} m_{crank} \tag{9}$$

$$P_{bat,MGP4} = \omega_{MGP4} T_{MGP4} + P_{loss,MGP4} \tag{10}$$

$$P_{bat,MGP0} = \omega_{MGP0} T_{MGP0} + P_{loss,MGP0} \tag{11}$$

where M_{fuel} is the total amount of fuel in grams consumed by the *ICE* up to time instant t, $\dot{m}_{fuel}$ is the fuel consumption rate in grams per second, m_{crank} is the mass of fuel required to crank the *ICE* (equal to 0.5 g here [33]), n_{crank} takes into account the number of activations over time, $P_{bat,MGP4}$ and $P_{bat,MGP0}$ are respectively the powers requested or supplied by the electric motors to the battery, whereas $P_{loss,MGP4}$ and $P_{loss,MGP0}$ indicate the power losses that characterize the two electric motors. These were evaluated by means of the lookup tables reported in Figure 2. The electrical, thermal, and ageing behavior of the battery could then be modeled as described in the following section.

3. High-Voltage Battery and Air Cooling System Modeling

The following methodologies were adopted here for modeling the plug-in HEV high-voltage battery pack from electrical, thermal, and ageing perspectives:

1. An equivalent circuit model for battery electrical modeling;

2. A single temperature lumped-parameter model that evaluates battery temperature evolution during HEV operation, which is coupled with the air cooling system model;
3. A throughput-based battery capacity fade model that evaluates battery cyclic ageing.
4. Each of the listed modeling methodologies is illustrated in the follow-up of this section.

3.1. Equivalent Circuit Model

Following a common approach from literature, the battery pack was modeled with an equivalent circuit [34]. To this end, Thevenin's theorem was applied to the pack configuration illustrated in Table 1 and constituted by A123 cells. In particular, the following equations hold:

$$OCV_{bat} = N_s OCV_{cell} \tag{12}$$

$$R_{bat} = \frac{N_s}{N_p} R_{cell} \tag{13}$$

where N_s and N_p are respectively the number of cells in series per branch and the number of branches in parallel, OCV_{cell} and OCV_{bat} represent open circuit voltages of one single cell and of the high-voltage battery pack, whereas R_{cell} and R_{bat} are the internal resistances of the cell and pack, respectively.

Similarly, cell and battery pack capacities C_{cell} and C_{bat} were linked as in Equation (14):

$$C_{bat} = N_p C_{cell}. \tag{14}$$

Battery current I_{bat} can be computed by means of Equation (15) by applying the power conservation law to the battery equivalent circuit model and by introducing battery power P_{bat}, which is detailed in Equations (22)–(25) in Section 4:

$$I_{bat} = \frac{OCV_{bat} - \sqrt{OCV^2{}_{bat} - 4P_{bat}R_{bat}}}{2R_{bat}}. \tag{15}$$

The instantaneous battery C-rate c could then be evaluated using Equation (16):

$$c = \frac{|I_{bat}|}{C_{bat}}. \tag{16}$$

Battery state-of-charge SOC could consequently be evaluated by knowing its value at the beginning of the driving mission (SOC_0) and using the current integration method:

$$SOC = SOC_0 - \int_0^t \frac{I_{bat}}{C_{bat}} dt. \tag{17}$$

It should be noted that cell parameters are not constant, but rather their values change during HEV operation. With respect to OCV_{cell}, its value varies with SOC, whereas R_{cell} depends on more factors, namely SOC, C rate, battery temperature, and charging/discharging conditions [14]. OCV_{cell} is sensitive to temperature variation as well, but its variation normally has less impact.

3.2. Single Temperature Lumped-Parameter Model and Battery Cooling System

Temperature has an extremely major impact on battery lifetime, which is especially holds when it falls beyond the ideal thermal thresholds that are currently estimated within 15 °C and 35 °C [12]. The high-voltage battery pack is an extended body characterized by a spatial distribution of temperature values, which is rather complex to model [35,36]. A wide range of battery thermal modeling options is available in the literature. Among these, lumped-parameter thermal models can be suggested as a good trade-off between model accuracy and ease of implementation in Matlab/Simulink® software. A single temperature lumped-parameter model of the battery has been implemented here by taking inspiration from the works by Han et al. [37] and Jaguemont et al. [38].

The battery pack considered in this paper is constituted by a total of 1440 cylindrical cells of type A123 26650. As reported by the manufacturer [26], each cell is shaped in a cylinder of Ø26 × 65 mm and weights of 76 g. Reasonable values for battery and air thermal parameters were obtained by combining cell manufacturer data with data extracted from the work presented by Kim et al. in 2013 [39]. These values have been reported in Table 2. In particular, they refer to the battery pack mass m_{bat}, to its specific heat c_{bat} and to its sizes: length l_{bat}, height H_{bat}, and width w_{bat}. S_{bat} represent the exchange surface with the surrounding air, which characterizes for a specific heat $c_{p,air}$. Despite being modeled using a thermal lumped approach, the battery pack was not assumed to be lumped from a spatial point of view, as displayed in Figure 3. Notably, 40, 3, and 12 cells were particularly aligned along l_{bat}, H_{bat}, and w_{bat}, respectively. The assembly of the battery pack was conceived to leave sufficient space for the channels in which the air flows when the cooling system is in operation. These channels were assumed to be 15 mm wide, and they constitute the overall cooling exchange surface between air and cells $S_{cooling}$ whose value is reported in Table 2. $h_{side,air}$ and $h_{cooling,air}$ are convective coefficients for the air surrounding the battery and the air forced to circulate in the cooling channels, respectively.

Table 2. Battery and air thermal parameters [26,39].

Parameter	Symbol	Value
Battery pack mass	m_{bat}	109.4 kg
Battery pack length	l_{bat}	1050 mm
Battery pack height	H_{bat}	240 mm
Battery pack width	w_{bat}	400 mm
Battery side exchange surface	S_{bat}	1.10 m^2
Overall cooling exchange surface	$S_{cooling}$	2.55 m^2
Battery specific heat	c_{bat}	1109.2 J/(kg·K)
Air specific heat	$c_{p,air}$	1005 J/(kg·K)
Convective coefficient—side air	$h_{side,air}$	10 W(m^2·K)
Convective coefficient—cooling air	$h_{cooling,air}$	50 W(m^2·K)

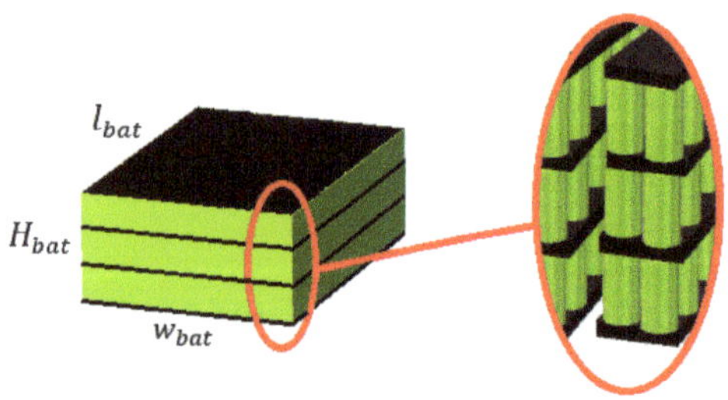

Figure 3. Schematic diagram of the arrangement of cells within the modeled battery pack.

The rest of this sub-section is structured as follows: the air cooling system for cabin and battery is described first. The numerical thermal modeling approach adopted for the battery pack is then illustrated, and the implemented HEV battery thermal management strategy is finally described.

3.2.1. Air Cooling System

The battery air cooling system has been assumed here to be capable of exploiting the cabin conditioned air, as suggested for example in the paper by Han et al. [37]. Air cooling is less effective than liquid cooling, however this approach is still widely used thanks to its limited cost and its ease of installation [40]. Moreover, several examples of its application can be found in the literature when aiming to evaluate the thermal behavior of cylindrical Li-ion cells [41,42]. Furthermore, the objective of this work does not relate to identifying the most suitable cooling system, but rather to analyze how the operation of a realistic cooling system impacts on the effective electrical power consumption during a given driving mission. A scheme of the cabin and battery air cooling system implemented here is reported in Figure 4.

Figure 4. Schematic diagram of the cabin and battery air cooling system.

When the battery cooling system is activated, air flow is enabled between vehicle cabin and high-voltage battery thanks to Fan2 in Figure 4. In this case, air flows through the battery in the dedicated channels between the cells. A forced convection heat exchange occurs in this framework through the surface $S_{cooling}$. Cabin climate comfort of the passengers might be slightly affected by such air flow circulation between the cabin and high voltage battery [40]. If cabin air conditioning is not activated by the HEV passengers, the air flow enabled by Fan2 to cool the battery will be at ambient temperature, thus considerably limiting the cooling effect. On the other hand, if cabin air conditioning is activated by the passengers, air flow will be enabled by Fan1 in Figure 4 from the *HVAC* (Heating, Ventilation, and Air Conditioning) system to the cabin. The air flow coming from the *HVAC* system will be at the temperature set by the HEV passengers for the air conditioning system. When the battery cooling system is not activated, the battery pack only dissipates heat with the surrounding air at cabin temperature by means of natural convection and passive cooling through its side surface S_{bat}.

3.2.2. Battery Thermal Model

The heat generated by the battery due to Joule effect (Q_{joule}) can take three different paths: a portion $Q_{bat,heat}$ remains stored in the battery and contributes to increasing its temperature, while a quantity $Q_{conv,nat}$ is dissipated by means of natural convective phenomena with the surrounding air. This latter is expected to remain at a constant temperature whose values refer to the cabin temperature T_{cab}. The final heat path $Q_{conv,\ AC}$ is different from zero only when the battery cooling system is in operation. This represents the portion of heat transferred from the battery pack to the air flowing at cabin temperature T_{cab} thanks to Fan2 being activated. When the *HVAC* system is not in operation, the cabin temperature T_{cab} has been considered as equal to the ambient temperature. On the other hand, when the HEV passengers activate the air conditioning, cabin temperature has been assumed to be constantly at 20 °C. The thermal balance for the battery pack can thus be obtained as in Equation (18):

$$Q_{joule} = Q_{bat,heat} + Q_{conv,nat} + Q_{conv,\ AC} \tag{18}$$

which in turn can be written by detailing each term as follows:

$$\begin{aligned} R_{bat} \cdot I_{bat}{}^2 = m_{bat} \cdot c_{bat} \cdot \tfrac{dT_{bat}}{dt} + h_{side,air} \cdot S_{bat} \cdot (T_{bat} - T_{cab}) + \\ h_{cooling,air} \cdot S_{cooling} \cdot (T_{bat} - T_{cab}) \cdot (Bat_{cooling} > 0) \end{aligned} \tag{19}$$

where m_{bat} , c_{bat}, T_{bat}, and S_{bat} refer respectively to battery mass, specific heat, temperature, and exchange surface with the surrounding air. $Bat_{cooling}$ is a binary flag and its value is set to 1 when the battery air cooling system is in operation (i.e., Fan2 is activated). All the thermal parameters have been reported in Table 2. It should be noted that Q_{joule} in

Equation (19) has been considered to be related to Joule's heating only. Battery temperature variation due to entropy changes was not considered here in the same way as normally assumed in the literature [43].

It should be admitted that a few assumptions were made for the illustrated battery thermal model. First, the formulation for $Q_{conv,\ AC}$ used in Equation (19) assumes that the air flow remains at a constant temperature equal to T_{cab} during the heat exchange with the battery pack. Moreover, assuming a constant value for the convective coefficient $h_{cooling,air}$ is not rigorous. Indeed, the convective coefficient is not an intrinsic property of the considered fluid, rather it also depends on the geometry of the object immersed in the fluid flow and on the flow rate magnitude [35]. As consequence, the evaluation of $h_{cooling,air}$ is rather complex and it is normally obtained by means of empirical correlations available in the literature. These correlations relate to forced convection and they consider the a-dimensional coefficients of Reynolds, Prandtl, and Nusselt [42]. However, these correlations normally refer to very simple cases and they might lead to important errors if applied to more complicate configurations. Although these approximations are not realistic, they aim to avoid introducing further modeling inaccuracy and they could therefore be retained for the purposes of this paper.

3.2.3. Battery Thermal Management Strategy

In this paper, a reactive battery thermal management strategy is considered. Inputs for the retained strategy include current values for battery temperature T_{bat} and battery air cooling system state $Bat_{cooling}$. Two threshold vales of battery temperature for the battery thermal management strategy were defined as respectively being related to $T_{bat-cooling,ON}$ and $T_{bat-cooling,OFF}$. The battery air cooling system was activated when the battery temperature increases above $T_{bat-cooling,ON}$. $Bat_{cooling}$ switches from 0 to 1 in this case. The battery air cooling system was set to operate until the battery temperature falls below $T_{bat-cooling,OFF}$. As is shown in Section 4, choosing the values for these two temperatures as function of the ambient temperature is not trivial and requires fine tuning.

3.3. *Throughput-Based Battery Capacity Fade Model*

A largely employed throughput-based capacity fade model was used to evaluate the high-voltage battery cyclic ageing. This model was calibrated for A123 26650 cells and it is based on the concept of battery charge throughput (Ah-throughput) [14]. Here, the Ah-throughput (Q_{EOL}) processed throughout the battery lifetime included the energy charged from the grid [16]. Q_{EOL} was associated to the overall quantity of charge (in ampere-hours) that a battery is able to process before reaching its end of life and it is calculable as:

$$Q_{EOL} = \int_0^{t_{EOL}} |I_{bat}| dt \tag{20}$$

where t_{EOL} is the time instant in which *EOL* occurs, which depends on the battery operating conditions. The *EOL* is considered to happen when the percentage value of battery nominal capacity faded (indicated by $C_{fade,\%}$) equals 20% (i.e., $C_{fade,\%} = 20$). When the battery operates under constant conditions, Q_{EOL} and $C_{fade,\%}$ could be evaluated using the equation proposed in [14] for A123 26650 cells. This formulation was inspired by the Arrhenius' gas equation [7] and it can only be employed for battery temperature ranging between 15 °C and 60 °C:

$$C_{fade,\%} = B \cdot exp\left(-\frac{A_f}{T_{bat}}\right)\left(\frac{Q_{EOL,const}}{N_p \cdot 1\ \text{Ah}}\right)^z \tag{21}$$

where battery Ah-throughput was indicated as $Q_{EOL,const}$ to specify that it refers to constant operating conditions. B, A_f, and z respectively indicate the pre-exponential, the activation, and the power-law factors. Both B and A_f are a function of the cell C-rate c [31]. Since these parameters refer to a single cell, $Q_{EOL,const}$ was divided by the number of parallel branches N_p in order to scale to the whole battery pack. Table 3 reports the values of parameters used in the considered battery ageing model. The values of pre-exponential factor B as function of the C-rate were determined by conducting an ageing experimental campaign [23].

Table 3. Battery ageing parameters for A123 26650 cell [23].

Parameter	Symbol	Value
C-rate index for pre-exponential factor	c	[0.5; 2; 4; 6; 8; 10; 12; 14; 16; 18; 20]
Pre-exponential factor as function of C-rate	$B(c)$	[31,630; 21,681; 17,307; 12,934; 13,512; 15,512; 12,099; 11,380; 13,656; 16,342; 14,599]
Activation factor	$A_f(c)$	3814.68–44.56·c
Power-law factor	z	0.55

As B and A_f vary according to the battery C-rate, $Q_{EOL,const}$ depends on both the operating conditions T_{bat} and C-rate, as illustrated in Figure 5. The range of the battery pack C-rate allowed as function of the considered plug-in HEV characteristics (i.e., battery pack capacity, *MGP*0 maximum power, and *MGP*4 maximum power) is highlighted in Figure 5. The larger capacity of a plug-in HEV battery slightly reduces the overall values of the C-rate, and this implies that the variation in $Q_{EOL,const}$ might largely relate to thermal characteristics. Looking again at Figure 5, thermal phenomena are forecasted to have a remarkable influence, even when the battery temperature is contained within the ideal range of 15 °C to 35 °C.

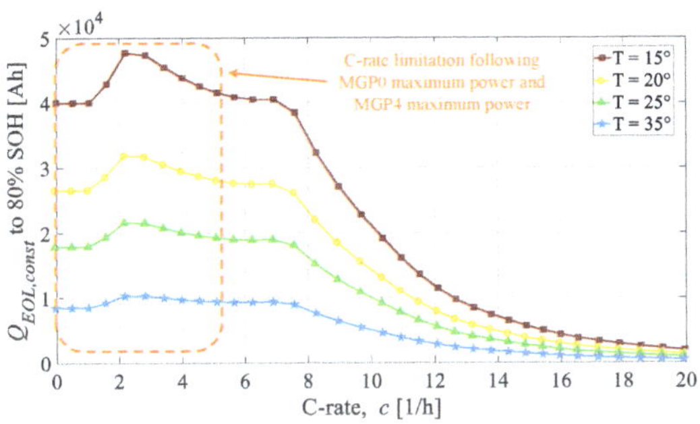

Figure 5. Allowed lifetime Ah-throughput predicted by the numerical ageing model as a function of the cell temperature and C-rate.

After calculating $Q_{EOL,const}$ using Equation (21), the overall quantity of charge Q_{const} supplied by the battery under fixed values of C-rate and temperature became known. Then, it became possible to evaluate the current battery ageing state by comparing Q_{const} with $Q_{EOL,const}$. However, since batteries employed in HEVs are characterized by varying operation over time, the contribution of each combination of the C-rate and T_{bat} towards the achievement of Q_{EOL} over time could be evaluated by integrating the instantaneous SOH variation:

$$SOH = SOH_0 - \int_0^t \frac{|I_{bat}|}{Q_{EOL,const}(c, T_{bat})} dt \tag{22}$$

where SOH_0 is the initial state-of-health of the battery and it is equal to 1 at the beginning of battery life. Predicting the battery SOH variation over time as a function of C-rate and temperature is possible in this way thanks to the implemented numerical ageing model.

4. Battery High Temperature Sensitive Optimization Based HEV Energy and Thermal Management Calibration

The optimization-driven calibration of the HEV on-line energy and thermal management strategies will be discussed in this section. Indeed, an adaptation of HEV energy and thermal management according to the ambient temperature value may allow finding the optimal trade-off between an enhanced fuel economy and satisfactory high-voltage battery lifetime. Table 4 summarizes variables and the calibration objective function for the optimization-based calibration of the HEV energy and thermal management. The calibration of HEV energy and thermal management is desired to be optimal and it must bring into play both $T_{bat-cooling,ON}$ and $T_{bat-cooling,OFF}$. Values for these parameters will depend both on the ambient temperature T_{amb} and on whether the *HVAC* system is in operation or not. Moreover, adapting the rule-based HEV EMS according to the battery temperature and predicted ageing rate may improve the accuracy and effectiveness of the vehicle control system. This work particularly considers varying the battery *SOC* which determines the transition from the charge depleting in pure electric (*EV*) mode to blended charge-depleting (HEV) hybrid as a function of T_{amb}. This value of battery *SOC* threshold was initially set to 0.30 for the baseline HEV energy management approach described in Section 2.1 and now it is indicated as $SOC_{EV,OFF}$.

Table 4. Variables and objective function for the optimization based HEV energy and thermal management calibration.

	Parameter	Symbol	Bounds
Calibration variables	Upper battery pack temperature threshold for cooling system activation	$T_{bat-cooling,ON}$ $(T_{amb}, HVAC_{state})$	$[T_{amb}–40°]$
	Lower battery pack temperature threshold for cooling system de-activation	$T_{bat-cooling,OFF}$ $(T_{amb}, HVAC_{state})$	$[10°–T_{amb}]$
	Lower battery *SOC* threshold for pure electric to hybrid electric operation transition	$SOC_{EV,OFF}$ $(T_{amb}, HVAC_{state})$	[0.3–1]
Calibration objective function	Operative cost associated to the entire plug-in HEV lifetime	$J_{calibration}$ (31)	

Figure 6 illustrates the flowchart for simulating the HEV in a given driving mission according to the implemented energy and thermal management approaches. The three parameters that need calibration in this framework ($T_{bat-cooling,ON}$, $T_{bat-cooling,OFF}$, $SOC_{EV,OFF}$) are highlighted in Figure 6. When the HEV driver starts the driving mission, the battery temperature was assumed to be equal to the ambient temperature. The *HVAC* system state was selected at the beginning of the driving mission, and it was considered to be constant throughout the journey. When the *HVAC* system was set to be in operation, the cabin temperature was assumed to be 20 °C from the first instant of the driving mission while neglecting transient thermal phenomena in the cabin.

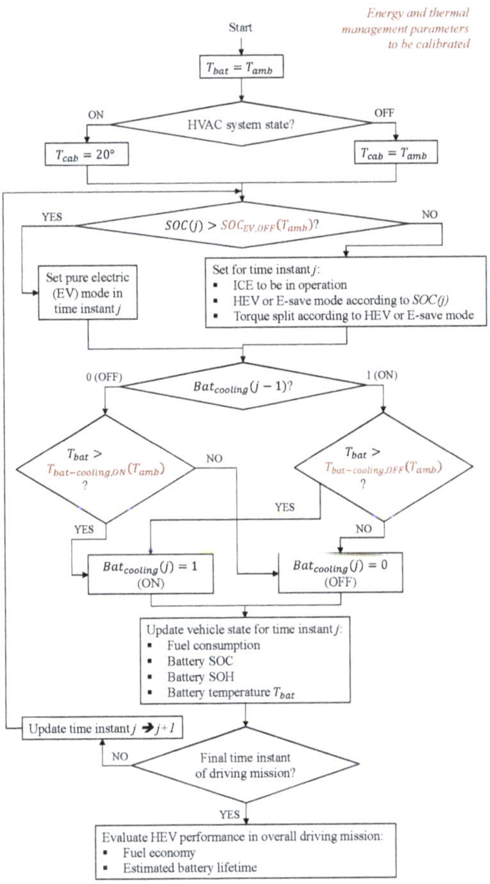

Figure 6. Flowchart for simulating the HEV in a driving mission according to the implemented energy and thermal management approaches.

At each time instant, the terms impacting on the instantaneous battery power P_{bat} used in Equation (15) were updated. The *HVAC* system state, the battery air conditioning system state, and the ambient temperature were considered to this end as in Equations (23)–(25):

$$P_{bat} = P_{bat,MGP0} + P_{bat,MGP4} + P_{aux} + P_{HVAC}(T_{amb}, HVAC_{state}) + P_{bat-cooling}\left(Bat_{cooling}\right) \quad (23)$$

$$P_{HVAC}(T_{amb}, HVAC_{state}) = 1000 + (T_{amb} - 20)^2 \cdot (HVAC_{state} > 0) \quad (24)$$

$$P_{bat-cooling}\left(Bat_{cooling}\right) = 200 \cdot \left(Bat_{cooling} > 0\right) \quad (25)$$

where P_{aux}, P_{HVAC}, and $P_{bat-cooling}$ stand for the power demand of the auxiliaries (e.g., lubrication, *ICE* cooling, lighting), the *HVAC* system electrical power, and the battery air conditioning system electrical power, respectively. A constant value of 400 W was set here for P_{aux}, while a value of the electrical load of the *HVAC* system was assumed in Equation (24) as a function of the ambient temperature by analyzing data presented in [44]. $HVAC_{state}$ represents a binary variable detecting *HVAC* system activation at the beginning of the driving mission. The power consumed by the battery air conditioning system ($P_{bat-cooling}$) mainly relates to Fan2 in Figure 4 and it was assumed here to be equal to 200 W [45].

4.1. HEV Fuel Economy and Battery Lifetime over Retained Driving Mix

An appropriate objective function needs definition to guide the optimization-based calibration of the HEV energy and thermal management parameters, i.e., $T_{bat-cooling,ON}$, $T_{bat-cooling,OFF}$, and $SOC_{EV,OFF}$. The minimization of HEV operative costs was considered in this work by simulating a realistic driving mix. Eight driving missions were retained including four standard drive cycles such as the worldwide harmonized light-vehicle test procedure (WLTP), the federal test procedure (FTP75), the RTS aggressive cycle (RTS95), and the highway federal test procedure (HWFET). Moreover, four real-world driving missions recorded by the authors by means of the global positioning system (GPS) were considered including up-hill (RWC01), 2.7-h highway (RWC02), down-mountain (RWC03), and downhill (RWC04) driving conditions. Time series for vehicle speed and net road altitude over time for all the eight driving missions retained are illustrated in Figure A1 in Appendix A, while Table 5 reports driving statistics. The overall vehicle lifetime in terms of mileage was divided while considering the eight driving missions steadily repeated. Each journey was associated with a coefficient α_{freq} in Table 5 that represents the percentage of vehicle lifetime which was assumed to be traveled in the corresponding driving mission.

Table 5. Statistics for retained driving missions.

Driving Mission	WLTP	FTP75	RTS95	HWFET	RWC01—Uphill	RWC02—Long Highway	RWC03—Down-Mountain	RWC04—Downhill
Length L_{cycle} [km]	23.3	17.8	12.9	16.5	17.8	296	27.4	16.7
Time [s]	1800	1878	887	765	1031	9792	2345	1123
Max speed [km/h]	131.3	91.2	134.4	96.3	112.7	135.4	84.9	103
Average speed [km/h]	46.5	34.1	52.5	77.6	62.1	108.8	42	53.5
Average acceleration [m/s^2]	0.4	0.5	0.8	0.2	0.5	0.2	0.5	0.6
Average deceleration [m/s^2]	−0.4	−0.6	−0.8	−0.2	−0.6	−0.2	−0.6	−0.6
Max altitude—min altitude [m]	0	0	0	0	278	523	682	225
Final altitude—initial altitude [m]	0	0	0	0	235	341	−633	−148
Journey frequency α_{freq} [%]	8.33	16.67	16.67	16.67	8.33	8.33	8.33	16.67

The predicted HEV fuel economy, HEV electrical energy economy, and battery lifetime over the entire vehicle lifetime considering the introduced driving mix could be expressed as:

$$Fuel_{avg} = \sum_{k=1}^{8} Fuel_k \cdot \alpha_{freq,k} \tag{26}$$

$$Electricity_{avg} = \sum_{k=1}^{8} Electricity_k \cdot \alpha_{freq,k} \tag{27}$$

$$batt_{life,avg} = \sum_{k=1}^{8} \frac{L_{cycle,k}}{SOH_0 - SOH_{end,k}} \cdot \alpha_{freq,k} \tag{28}$$

where $Fuel_k$ and $Fuel_{avg}$ are the HEV fuel economy values expressed in liters per 100 km related to the given driving mission k and the entire vehicle lifetime, respectively. $Electricity_k$ and $Electricity_{avg}$ are the plug-in HEV electrical energy economy values expressed in kilowatt-hours per 100 km related to the given driving mission k and the entire vehicle lifetime, respectively. $L_{cycle,k}$ stand for the mileage in kilometers of the given driving mission k, while $SOH_{end,k}$ is the battery SOH at the end of the driving mission k according to the HEV simulated following the discussed energy and thermal management strategies.

4.2. Workflow for Optimization-Based HEV Thermal and Energy Management Calibration

This sub-section aims at introducing the flowchart of the optimization-based calibration of $T_{bat-cooling,ON}$, $T_{bat-cooling,OFF}$, and $SOC_{EV,OFF}$ as function of the ambient temperature. When establishing the energy and thermal management calibration objective, minimizing the operative cost related to the entire lifetime of the plug-in HEV was considered. Retained vehicle cost contributions particularly include:

- The HEV fuel consumption;
- The electrical energy consumed by the plug-in HEV coming from the grid;
- The monetary cost associated with the battery pack degradation.

In this work, particle swarm optimization (PSO) was selected as an algorithm for exploring the identified three-dimensional calibration space with $T_{bat-cooling,ON}$, $T_{bat-cooling,OFF}$, and $SOC_{EV,OFF}$ as independent variables. PSO is an iterative stochastic optimization algorithm that relies on a numerical model capturing the social behavior of fishes and birds proposed by Kennedy and Eberhart in 1995 [46]. The flowchart of the PSO approach implemented in this work is illustrated in Figure 7 and detailed as follows. During step 1, a population of N particles was initialized aiming at the subsequent exploration of the calibration space searching for the minimum of the PSO cost function $J_{calibration}$. For each value of the selected ambient temperature T_{amb} and $HVAC$ system state, at each iteration of the PSO algorithm, step 2 aims at assessing the value of $J_{calibration}$ for each particle of the swarm. This was performed by simulating the HEV traveling through all the eight retained driving missions constantly in Simulink®® software at the selected value of ambient temperature and considering both the selected $HVAC$ system state being constant and the battery being fully charged at the beginning of the driving mission (i.e., the SOC was set to 95% as the maximum value assumed). For the sake of an exhaustive analysis, two different vehicle loading conditions were considered when simulating each driving mission. These relate to consider either the vehicle lightly loaded (i.e., $n_{pass} = 1$) or fully loaded (i.e., $n_{pass} = 5$), respectively. As all the simulations of the driving missions were completed, the fuel economy, electrical energy economy and battery lifetime weighted according to both driving mix and vehicle payload conditions could be computed as:

$$Fuel_{avg,total} = \alpha_{payload} \cdot Fuel_{avg,LLV} + \left(1 - \alpha_{payload}\right) \cdot Fuel_{avg,FLV} \quad (29)$$

$$Electricity_{avg,total} = \alpha_{payload} \cdot Electricity_{avg,LLV} + \left(1 - \alpha_{payload}\right) \cdot Electricity_{avg,FLV} \quad (30)$$

$$batt_{life,total} = \alpha_{payload} \cdot batt_{life,avg,LLV} + \left(1 - \alpha_{payload}\right) \cdot batt_{life,avg,FLV} \quad (31)$$

where $Fuel_{avg,total}$, $Electricity_{avg,total}$, and $batt_{life,total}$ respectively represent HEV fuel economy in litres per 100 km, electrical energy economy in kilowatt-hours per 100 km, and kilometrical battery lifetime weighted according to both driving mix and payload conditions. Subscripts LLV and FLV were associated with lightly loaded and fully loaded vehicle conditions, respectively. Finally, $\alpha_{payload}$ represents a weighting coefficient for the two loading conditions. Its value was set here to 0.0825 following a press release by the association Anci suggesting that, in Italy, the average number of passengers per car is equal to 1.33 [47].

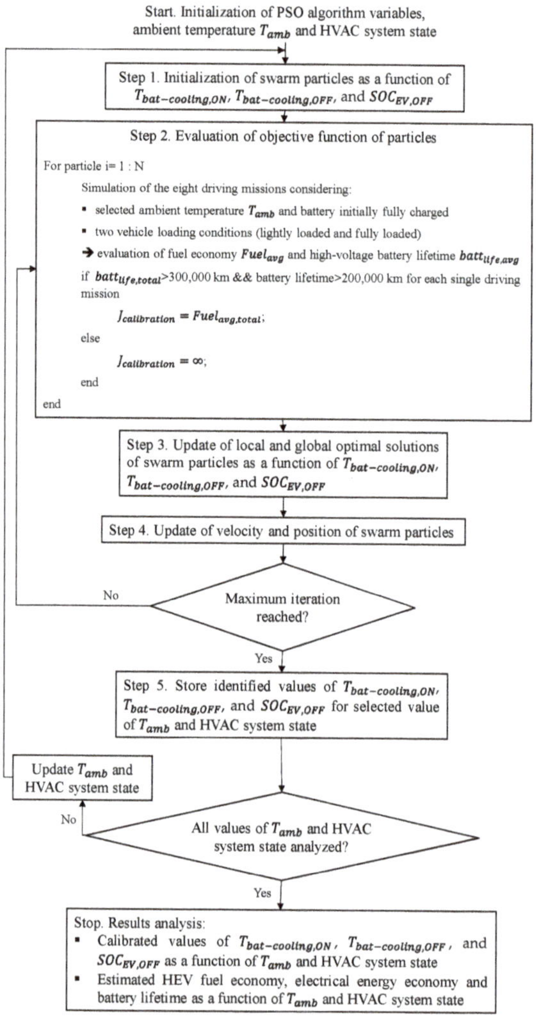

Figure 7. Flowchart of the particle swarm optimization-based calibration of the parameters for HEV energy and thermal management.

The cost function $J_{calibration}$ associated to the particle under analysis could then be evaluated including monetary costs associated with fuel consumption, electrical energy consumption, and battery degradation for the target plug-in HEV lifetime in kilometres $veh_{life,target}$, which was assumed to be 300 thousand km here [48]. The resulting mathematical formulation for $J_{calibration}$ was as follows:

$$J_{calibration} = c_{fuel} \cdot \frac{Fuel_{avg,total}}{100} \cdot veh_{life,target} + c_{elec} \cdot \frac{Electricity_{avg,total}}{100} \cdot veh_{life,target} + c_{batt} \cdot \frac{veh_{life,target}}{batt_{life,total}} \quad (32)$$

where c_{fuel}, c_{elec}, and c_{batt} respectively represent the fuel refuelling cost, the cost of the electricity coming from the grid, and the battery pack replacement cost. Values of c_{fuel} and c_{elec} retained here correspond to 1.41 euros per liter and 0.22 euros per kilowatt-hour, which relate to average prices observed in Italy in the first half of 2020 [49,50]. Concerning c_{batt}, its value was assumed here to be around 6130 euros, considering the current estimated battery pack specific production cost of 280 euros per kilowatt-hour and multiplying it by the

HEV battery pack capacity and a factor of 2, which accounts for manpower and other costs associated with the battery pack replacement process [51]. The battery pack replacement cost term was not considered here in case $batt_{life,total}$ exceeded the target value for the entire vehicle lifetime $veh_{life,target}$. $J_{calibration}$ thus considers the operative costs in euros associated with the entire plug-in HEV lifetime involving fuel consumption, electricity consumption, and battery pack degradation. Once $J_{calibration}$ was evaluated for all the particles of the swarm, step 3 in Figure 7 aims to update the local and global minima for the swarm. Position and velocity of each particle in the three-dimensional calibration space considered were then updated in step 4 based on the local and global optima previously identified. Three main PSO parameters were involved in this procedure, namely the inertia parameter, the cognitive parameter, and the social parameter [52]. More details regarding this step can be found in [53].

Once the maximum number of iterations was reached for the PSO algorithm, the obtained results included the calibrated values of $T_{bat-cooling,ON}$, $T_{bat-cooling,OFF}$, and $SOC_{EV,OFF}$ for the selected value of ambient temperature and *HVAC* system state. Compared with other popular stochastic optimization algorithms (e.g., genetic algorithm—GA), PSO distinguished itself by the ease of its management and parameter tuning [54,55]. This corroborates the likelihood of effectively finding a global optimum for the considered calibration problem and it represents the main motivation behind the use of PSO in this work. The version of the PSO algorithm considered in this paper refers to the corresponding toolbox provided by the Yarpiz project and it was implemented in MATLAB© software [56].

In the next section, results are provided for the energy and thermal HEV management strategies calibrated according to the illustrated methodology.

5. Results

In this work, the HEV energy and thermal management strategies discussed earlier find calibration in terms of $T_{bat-cooling,ON}$, $T_{bat-cooling,OFF}$, and $SOC_{EV,OFF}$ according to eight different values of ambient temperature corresponding to 15, 18, 21, 24, 27, 30, 33, and 36 °C. Retained parameters for the PSO-based calibration flowchart are reported in Table 6 [56].

Table 6. Retained PSO parameters.

Parameter	Value
Swarm size [-]	20
Number of iterations [-]	15
Inertia coefficient [-]	0.73
Cognitive coefficient [-]	1.5
Social coefficient [-]	1.5

Considering a desktop computer with Intel Core i7-8700 (3.2 GHz) and 32 GB of RAM, around 320 min were required to complete step 1 to step 5 of Figure 7. In total, 16 iterations of the workflow from step 1 to step 5 were then required when considering eight values of ambient temperature and two *HVAC* system states (i.e., on or off). This led to around 43 h of computational time being required for completing the entire HEV energy and thermal management calibration procedure illustrated. Figure 8 illustrates the values of $T_{bat-cooling,ON}$, $T_{bat-cooling,OFF}$, and $SOC_{EV,OFF}$ obtained after the completion of the overall PSO based calibration workflow illustrated in Figure 7. Concerning the thermal management parameters shown in Figure 8a, a large gap between $T_{bat-cooling,ON}$ and $T_{bat-cooling,OFF}$ meant that the battery air conditioning system could be kept operating for longer after being activated. On the other hand, a narrower difference between the values of $T_{bat-cooling,ON}$ and $T_{bat-cooling,OFF}$ entailed more frequent activations and deactivations of the battery air conditioning system. When the *HVAC* system was off, a larger gap between $T_{bat-cooling,ON}$ and $T_{bat-cooling,OFF}$ was observed below 30 °C, which involves more usage of the battery air conditioning system. However, this correlated to a substantial increase in the calibrated values for $SOC_{EV,OFF}$ in Figure 8b from around 0.33 to around 0.58 and progressively above as the ambient temperature overcomes 30 °C, especially for the *HVAC* system being de-activated. This means that, as the ambient temperature gradually increases, the PSO based calibration procedure prefers switching from pure electric operation to hybrid electric operation much earlier in a driving mission rather than retaining a pure electric operation while keeping the battery air conditioning system in operation for a shorter time when the *HVAC* system is off. The trend of $SOC_{EV,OFF}$ as function of the ambient temperature when the *HVAC* system was in operation showed a similar trend, yet slightly shifted downwards, with respect to the *HVAC* system being off in Figure 8b. As a consequence, the *HVAC* system being activated seemed to enable exploiting pure electric charge-depleting operation for longer at high ambient temperatures thanks to the availability of the cooled cabin air to cool down the battery pack. This correlates with a calibrated larger gap between $T_{bat-cooling,ON}$ and $T_{bat-cooling,OFF}$ for high ambient temperatures when the *HVAC* system is in operation in Figure 8a, which entails longer usage of the battery cooling system.

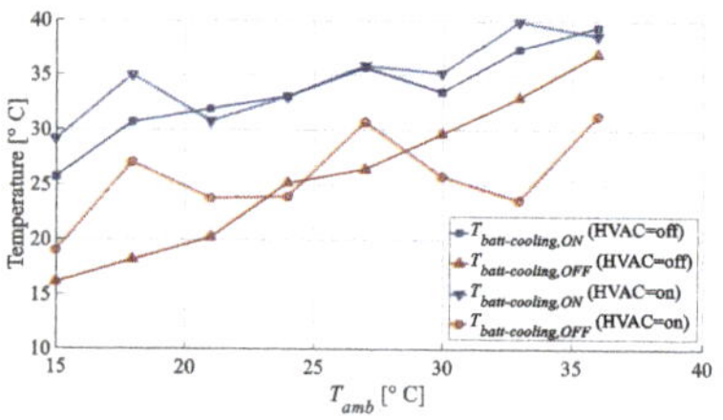

(a) Thermal management parameters

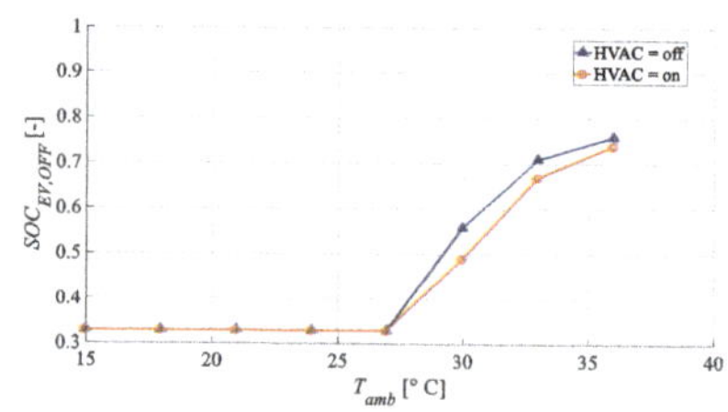

(b) Energy management parameter

Figure 8. Calibrated values of $T_{bat-cooling,ON}$, $T_{bat-cooling,OFF}$, and $SOC_{EV,OFF}$ according to the ambient temperature and *HVAC* system state.

Table 7 reports numerical values for the calibrated thermal and energy management parameters as a function of the ambient temperature and the *HVAC* system state. Values of $Fuel_{avg,total}$, $Electricity_{avg,total}$, and $batt_{life,total}$ are included along with the corresponding value of $J_{calibration}$. Moreover, a graphical representation of the considered plug-in HEV energy consumption as function of the ambient temperature is shown in Figure 9. As can be seen in Figure 9, fuel consumption and battery energy consumption are quite stable within the ambient temperature range of 15 °C to 30 °C thanks to the implemented calibration process for the HEV energy and thermal management strategy. As the value of $SOC_{EV,OFF}$ shown in Figure 8b was increased to preserve battery lifetime at around 33 °C, the corresponding battery energy consumption lowered by 11.2% and the corresponding fuel consumption increased by around two times compared to the 15 °C reference values in Figure 9a. The overall plug-in HEV energy consumption shown in Figure 9b considers the sum of battery electrical energy and fuel chemical energy in kilowatt-hours per 100 km. The fuel chemical energy could be obtained by considering values of fuel density and a lower heating value corresponding to 744 g per liter and 43.74 kilojoules/gram, respectively. In this case, a 33 °C ambient temperature might lead to 1.4 times the overall plug-in HEV energy consumption compared with 15 °C. As the ambient temperature further increased, the average fuel consumption increased, and the battery energy consumption reduced to prevent accelerated lifetime consumption. From the overall energy consumption perspective, 36 °C involved more than two times the energy consumption estimated at 15 °C in Figure 9b. This reflects operative costs in the increased values for the overall plug-in HEV at high ambient temperatures as well.

Table 7. Calibrated thermal and energy management parameters and HEV performance in terms of energy consumption and predicted battery lifetime as a function of the ambient temperature and *HVAC* system state.

HVAC System State	T_{amb} [°C]	$T_{bat-cooling,ON}$ [°C]	$T_{bat-cooling,OFF}$ [°C]	$SOC_{EV,OFF}$ [%]	$Fuel_{avg,total}$ [L/100 km]	$Electricity_{avg,total}$ [kWh/100 km]	$batt_{life,total}$ [km·10³]	$J_{calibration}$ [k€]
Off	15	25.7	16.1	33.4	0.69	16.57	799	13.8
Off	18	30.7	18.2	33.5	0.69	16.42	658	13.8
Off	21	31.9	20.2	33.4	0.69	16.41	521	13.8
Off	24	33.0	25.2	33.4	0.69	16.40	414	13.8
Off	27	35.7	26.4	33.5	0.69	16.39	331	13.8
Off	30	33.5	29.6	55.8	0.74	16.21	300	14.0
Off	33	37.4	32.9	70.7	1.44	14.21	300	15.6
Off	36	39.4	36.9	76.5	2.65	10.74	300	18.4
On	15	29.2	19.0	33.2	0.69	16.55	781	13.8
On	18	35.0	27.0	33.4	0.69	16.42	651	13.8
On	21	30.8	23.8	33.4	0.69	16.41	524	13.8
On	24	33.0	23.9	33.5	0.69	16.40	422	13.8
On	27	35.9	30.7	33.3	0.69	16.39	342	13.8
On	30	35.2	25.8	48.9	0.70	16.34	300	13.9
On	33	40.0	23.6	67.3	1.10	15.20	300	14.8
On	36	38.7	31.3	74.3	2.20	12.06	304	17.4

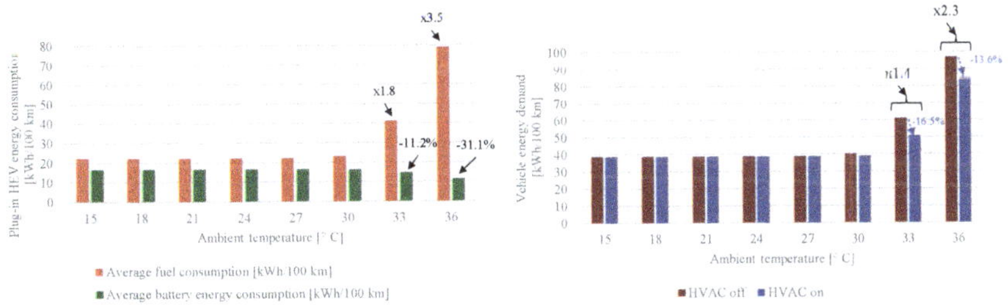

(a) $Fuel_{avg,total}$ and $Electricity_{avg,total}$ as a function of ambient temperature (average between HVAC system off and HVAC system on)

(b) Overall plug-in HEV energy demand considering both fuel chemical energy and battery electrical energy as a function of ambient temperature

Figure 9. Plug-in HEV fuel consumption, battery energy consumption, and overall energy consumption as a function of the ambient temperature while benchmarking with corresponding values at 15 °C.

The activation of the *HVAC* system impacted on the cabin temperature and, in turn, allowed cooling the battery pack with air at lower temperature compared with the ambient temperature. Nevertheless, a considerable amount of additional auxiliary power consumption was associated to the *HVAC* system activation. In this framework, Figure 9b suggests that, from the overall vehicle perspective, the beneficial effect of lowering the temperature of the battery air conditioning system overcame the increased power consumption when the *HVAC* system was activated. As a result, the *HVAC* system being in operation entailed a lower overall plug-in HEV energy consumption by 16.5% and 13.6% for ambient temperature equal to 33 °C and 36 °C, respectively.

Finally, it should be noted that values of $batt_{life,total}$ reported in Table 7 are always equal to or are higher than the set vehicle lifetime target of 300,000 km. The calibration performed by PSO thus suggests that, from the point of view of the plug-in HEV operative cost, slightly increasing the usage of *ICE* might be advised at high ambient temperatures in order to preserve battery lifetime and thus avoid the costs required for replacing the battery pack within the vehicle lifetime.

6. Conclusions

This paper proposes the calibration of the thermal and energy management strategies of a plug-in HEV to minimize the operative costs associated to fuel consumption, electricity consumption, and battery pack degradation for the overall vehicle lifetime at various ambient temperatures. A parallel though-the-road plug-in HEV powertrain architecture was modeled in Matlab/Simulink® software. Moreover, the battery pack was modeled from electrical, ageing, and thermal perspectives. A methodology based on PSO was proposed to calibrate three parameters of the HEV thermal and energy management aiming at minimizing overall vehicle operative cost. The exhaustiveness of the calibration procedure was improved by considering different ambient temperature, vehicle payload conditions, and *HVAC* system states.

Obtained results suggest that, as the ambient temperature increases, different settings of the battery air conditioning system should be implemented in the on-board HEV controller. Moreover, switching from a pure electric to hybrid electric operation below a certain value of the battery *SOC* can be advised to preserve battery lifetime and to avoid replacing the battery pack within the vehicle lifetime. High ambient temperatures impact on the overall plug-in HEV energy consumption to the point that at 36 °C, the HEV depletes 2.3 times the corresponding energy depleted at 15 °C. Results also suggested that the activation of the *HVAC* system for cabin conditioning at high ambient temperatures might have a beneficial impact on cooling the battery, which overcomes the increase in auxiliary power consumption due to the *HVAC* system.

Implementing the methodology introduced here can decrease time and engineering effort required to calibrate energy and thermal management strategies of HEVs. The satisfaction of conflicting calibration objectives can be enhanced in this way, including fuel economy and electrical energy economy enhancement, high-voltage battery lifetime safeguarding, and temperature control. Related future work might involve improving the fidelity level of both the HEV powertrain model, the battery pack thermal model, and the battery ageing model. Moreover, different design solutions in terms of cell chemistry and battery pack conditioning system could be investigated. Developing more sophisticated and accurate thermal and energy management might lead to further enhancing the overall energy economy of the HEV. Finally, performing experimental tests might prove fundamental to validate the proposed numerical methodology and the calibrated HEV thermal and energy management strategy.

Author Contributions: Conceptualization, P.G.A.; methodology, P.G.A. and M.D.P.; software, M.D.P.; validation, M.D.P. and P.G.A.; formal analysis, M.D.P., P.G.A., and G.B.; investigation, M.D.P. and P.G.A.; resources, G.B.; data curation, M.D.P. and P.G.A.; writing—original draft preparation, M.D.P. and P.G.A.; writing—review and editing, G.B.; visualization, M.D.P. and P.G.A.; supervision, P.G.A. and G.B.; project administration, P.G.A. and G.B.; funding acquisition, G.B. All authors have read and agreed to the published version of the manuscript.

Funding: This research received no external funding.

Data Availability Statement: No applicable.

Conflicts of Interest: The authors declare no conflict of interest.

Abbreviations

ECMS	Equivalent consumption minimization strategy
EMS	Energy management strategy
EOL	End-of-life
FTP75	Federal test procedure 75
HEV	Hybrid electric vehicle
HVAC	Heating, Ventilation, and Air Conditioning
HWFET	Highway federal test procedure
ICE	Internal combustion engine
MGP0	Motor/generator P0
MGP4	Motor/generator P4
OOL	Optimal operating line
PID	Proportional–integral–derivative
PSO	Particle swarm optimization
SEI	Solid electrolyte interface

SOC	State of charge
SOH	State of health
WLTP	Worldwide harmonized light-vehicle test procedure

Appendix A. Time Series of Retained Driving Missions

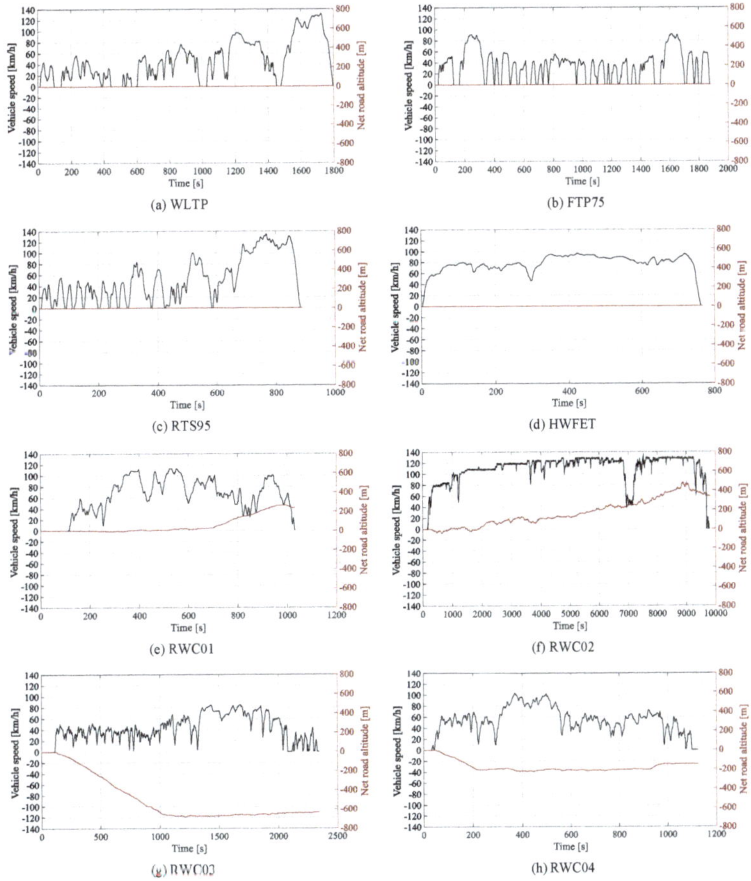

Figure A1. Vehicle speed and net road altitude profiles over time for the retained driving missions.

References

1. Bilgin, B.; Magne, P.; Malysz, P.; Yang, Y.; Pantelic, V.; Preindl, M.; Korobkine, A.; Jiang, W.; Lawford, M.; Emadi, A. Making the Case for Electrified Transportation. *IEEE Trans. Transp. Electrif.* **2015**, *1*, 4–17. [CrossRef]
2. Tran, M.-K.; Akinsanya, M., Panchal, S.; Fraser, R.; Fowler, M. Design of a Hybrid Electric Vehicle Powertrain for Performance Optimization Considering Various Powertrain Components and Configurations. *Vehicles* **2020**, *3*, 20–32. [CrossRef]
3. Emadi, A. *Advanced Electric Drive Vehicles*; CRC Press: Boca Raton, FL, USA, 2014.
4. Anselma, P.G.; Belingardi, G. *Next Generation HEV Powertrain Design Tools: Roadmap and Challenges*; SAE Technical Paper 2019-01-2602; SAE International: Warrendale, PA, USA, 2019.
5. Kollmeyer, P.J.; Jahns, T.M. Aging and performance comparison of absorbed glass matte, enhanced flooded, PbC, NiZn, and LiFePO4 12V start stop vehicle batteries. *J. Power Sources* **2019**, *441*, 227139. [CrossRef]
6. Anselma, P.G.; Kollmeyer, P.J.; Feraco, S.; Bonfitto, A.; Belingardi, G.; Emadi, A.; Amati, N.; Tonoli, A. Assessing Impact of Heavily Aged Batteries on Hybrid Electric Vehicle Fuel Economy and Drivability. In Proceedings of the 2021 IEEE Transportation Electrification Conference and Expo, Chicago, IL, USA, 21–25 June 2021.

7. Bloom, I.; Cole, B.; Sohn, J.; Jones, S.; Polzin, E.; Battaglia, V.; Henriksen, G.; Motloch, C.; Richardson, R.; Unkelhaeuser, T.; et al. An accelerated calendar and cycle life study of Li-ion cells. *J. Power Sources* **2001**, *101*, 238–247. [CrossRef]
8. Bonfitto, A. A Method for the Combined Estimation of Battery State of Charge and State of Health Based on Artificial Neural Networks. *Energies* **2020**, *13*, 2548. [CrossRef]
9. Vidal, C.; Kollmeyer, P.; Naguib, M.; Malysz, P.; Gross, O.; Emadi, A. Robust xEV Battery State-of-Charge Estimator Design Using a Feed-forward Deep Neural Network. *SAE Int. J. Adv. Curr. Prac. Mobil.* **2020**, *2*, 2872–2880.
10. Bonfitto, A.; Feraco, S.; Tonoli, A.; Amati, N.; Monti, F. Estimation Accuracy and Computational Cost Analysis of Artificial Neural Networks for State of Charge Estimation in Lithium Batteries. *Batteries* **2019**, *5*, 47. [CrossRef]
11. Bonfitto, A.; Ezemobi, E.; Amati, N.; Feraco, S.; Tonoli, A.; Hegde, S. State of Health Estimation of Lithium Batteries for Automotive Applications with Artificial Neural Networks. In Proceedings of the 2019 AEIT International Conference of Electrical and Electronic Technologies for Automotive (AEIT AUTOMOTIVE), Turin, Italy, 2–4 July 2019; pp. 1–5.
12. Vidal, C.; Gross, O.; Gu, R.; Kollmeyer, P.; Emadi, A. xEV Li-Ion Battery Low-Temperature Effects—Review. *IEEE Trans. Veh. Technol.* **2019**, *68*, 4560–4572. [CrossRef]
13. Rauhala, T.; Jalkanen, K.; Romann, T.; Lust, E.; Omar, N.; Kallio, T. Low-temperature aging mechanisms of commercial graphite/LiFePO4 cells cycled with a simulated electric vehicle load profile—A post-mortem study. *J. Energy Storage* **2018**, *20*, 344–356. [CrossRef]
14. Wang, J.; Liu, P.; Hicks-Garner, J.; Sherman, E.; Soukiazian, S.; Verbrugge, M.; Tataria, H.; Musser, J.; Finamore, P. Cycle-life model for graphite- $LiFePO_4$ cells. *J. Power Sources* **2011**, *196*, 3942–3948. [CrossRef]
15. Ebbesen, S.; Elbert, P.; Guzzella, L. Battery State-of-Health Perceptive Energy Management for Hybrid Electric Vehicles. *IEEE Trans. Veh. Technol.* **2012**, *61*, 2893–2900. [CrossRef]
16. Onori, S.; Spagnol, P.; Marano, V.; Guezennec, Y.; Rizzoni, G. A new life estimation method for lithium-ion batteries in plug-in hybrid electric vehicles applications. *Int. J. Power Electron.* **2012**, *4*, 302. [CrossRef]
17. Tang, L.; Rizzoni, G.; Onori, S. Energy Management Strategy for HEVs Including Battery Life Optimization. *IEEE Trans. Transp. Electrif.* **2015**, *1*, 211–222. [CrossRef]
18. Zhang, S.; Hu, X.; Xie, S.; Song, Z.; Hu, L.; Hou, C. Adaptively coordinated optimization of battery aging and energy management in plug-in hybrid electric buses. *Appl. Energy* **2019**, *256*, 113891. [CrossRef]
19. Xie, S.; Hu, X.; Zhang, Q.; Lin, X.; Mu, B.; Ji, H. Aging-aware co-optimization of battery size, depth of discharge, and energy management for plug-in hybrid electric vehicles. *J. Power Sources* **2020**, *450*, 227638. [CrossRef]
20. Cheng, M.; Chen, B. Nonlinear Model Predictive Control of a Power-Split Hybrid Electric Vehicle with Consideration of Battery Aging. *J. Dyn. Syst. Meas. Control.* **2019**, *141*, 081008. [CrossRef]
21. Anselma, P.G.; Kollmeyer, P.; Belingardi, G.; Emadi, A. Multi-Objective Hybrid Electric Vehicle Control for Maximizing Fuel Economy and Battery Lifetime. In Proceedings of the 2020 IEEE Transportation Electrification Conference & Expo (ITEC), Chicago, IL, USA, 22–26 June 2020; pp. 1–6.
22. Patil, C.; Naghshtabrizi, P.; Verma, R.; Tang, Z.; Smith, K.; Shi, Y. Optimal battery utilization over lifetime for parallel hybrid electric vehicle to maximize fuel economy. In Proceedings of the 2016 American Control Conference (ACC), Boston, MA, USA, 6–8 July 2016; pp. 1524–1529.
23. Anselma, P.G.; Kollmeyer, P.; Lempert, J.; Zhao, Z.; Belingardi, G.; Emadi, A. Battery state-of-health sensitive energy management of hybrid electric vehicles: Lifetime prediction and ageing experimental validation. *Appl. Energy* **2021**, *285*, 116440. [CrossRef]
24. Du, R.; Hu, X.; Xie, S.; Hu, L.; Zhang, Z.; Lin, X. Battery aging- and temperature-aware predictive energy management for hybrid electric vehicles. *J. Power Sources* **2020**, *473*, 228568. [CrossRef]
25. Fiat Chrysler Automobiles. Renegade 4xe and Compass 4xe: The Jeep®® Brand's Take on the Plug-in Hybrid. Available online: http://www.media.fcaemea.com/em-en/jeep/press/renegade-4xe-and-compass-4xe-the-jeep-brand-s-take-on-the-plug-in-hybrid (accessed on 28 May 2021).
26. A123 Systems. Nanophosphate®® High Power Lithium Ion Cell ANR26650M1-B. Available online: https://www.batteryspace.com/prod-specs/6610.pdf (accessed on 28 May 2021).
27. Uddin, K.; Perera, S.; Widanage, W.D.; Somerville, L.; Marco, J. Characterising Lithium-Ion Battery Degradation through the Identification and Tracking of Electrochemical Battery Model Parameters. *Batteries* **2016**, *2*, 13. [CrossRef]
28. Rizzoni, G.; Guzzella, L.; Baumann, B. Unified modeling of hybrid electric vehicle drivetrains. *IEEE/ASME Trans. Mechatron.* **1999**, *4*, 246–257. [CrossRef]
29. Alix, G.; Dabadie, J.-C.; Font, G. An *ICE* Map Generation Tool Applied to the Evaluation of the Impact of Downsizing on Hybrid Vehicle Consumption. *SAE Tech. Pap.* **2015**, *24*, 2385. [CrossRef]
30. Le Berr, F.; Abdelli, A.; Postariu, D.-M.; Benlamine, R. Design and Optimization of Future Hybrid and Electric Propulsion Systems: An Advanced Tool Integrated in a Complete Workflow to Study Electric Devices. *Oil Gas Sci. Technol.-Rev. d'IFP Energ. Nouv.* **2012**, *67*, 547–562. [CrossRef]
31. Anselma, P.G.; Kollmeyer, P.; Belingardi, G.; Emadi, A. Multitarget Evaluation of Hybrid Electric Vehicle Powertrain Architectures Considering Fuel Economy and Battery Lifetime. *SAE Tech. Pap.* **2020**, *37*, 15. [CrossRef]
32. United States Environmental Protection Agency. Compliance and Fuel Economy Data-Annual Certification Data for Vehicles, Engines, and Equipment. Available online: https://www.epa.gov/compliance-and-fuel-economy-data/annual-certification-data-vehicles-engines-and-equipment (accessed on 1 June 2021).

33. Engbroks, L.; Knappe, P.; Goerke, D.; Schmiedler, S.; Goedecke, T.; Geringer, B. Energetic Costs of ICE Starts in (P)HEV—Experimental Evaluation and Its Influence on Optimization Based Energy Management Strategies. *SAE Tech. Pap.* **2019**, *24*, 0203.
34. Tran, M.-K.; Mevawala, A.; Panchal, S.; Raahemifar, K.; Fowler, M.; Fraser, R. Effect of integrating the hysteresis component to the equivalent circuit model of Lithium-ion battery for dynamic and non-dynamic applications. *J. Energy Storage* **2020**, *32*, 101785. [CrossRef]
35. Maheshwari, A.; Dumitrescu, M.A.; Destro, M.; Santarelli, M. A modeling approach to understand charge discharge differences in thermal behavior in lithium iron phosphate–graphite battery. *Electrochim. Acta* **2017**, *243*, 129–141. [CrossRef]
36. Mevawalla, A.; Panchal, S.; Tran, M.-K.; Fowler, M.; Fraser, R. One dimensional fast computational partial differential model for heat transfer in lithium-ion batteries. *J. Energy Storage* **2021**, *37*, 102471. [CrossRef]
37. Han, T.; Khalighi, B.; Yen, E.C.; Kaushik, S. Li-Ion Battery Pack Thermal Management: Liquid Versus Air Cooling. *J. Therm. Sci. Eng. Appl.* **2018**, *11*, 021009. [CrossRef]
38. Jaguemont, J.; Boulon, L.; Dube, Y.; Martel, F. Thermal Management of a Hybrid Electric Vehicle in Cold Weather. *IEEE Trans. Energy Convers.* **2016**, *31*, 1110–1120. [CrossRef]
39. Kim, Y.; Mohan, S.; Siegel, J.B.; Stefanopoulou, A.G.; Ding, Y. The estimation of radial temperature distribution in cylindrical battery cells under unknown cooling conditions. In Proceedings of the 52nd IEEE Conference on Decision and Control, Florence, Italy, 1 June 2013.
40. Janarthanam, S.; Burrows, N.; Boddakayala, B.R. Factors Influencing Liquid over Air Cooling of High Voltage Battery Packs in an Electrified Vehicle. *SAE Tech. Pap.* **2017**, *1*, 1171. [CrossRef]
41. Lin, X.; Perez, H.E.; Mohan, S.; Siegel, J.; Stefanopoulou, A.G.; Ding, Y.; Castanier, M.P. A lumped-parameter electro-thermal model for cylindrical batteries. *J. Power Sources* **2014**, *257*, 1–11. [CrossRef]
42. Jeon, D.H.; Baek, S.M. Thermal modeling of cylindrical lithium ion battery during discharge cycle. *Energy Convers. Manag.* **2011**, *52*, 2973–2981. [CrossRef]
43. Lempert, J.; Kollmeyer, P.; Malysz, P.; Gross, O.; Cotton, J.; Emadi, A. Battery Entropic Heating Coefficient Testing and Use in Cell-Level Loss Modeling for Extreme Fast Charging. *SAE Int. J. Adv. Curr. Pract. Mobil.* **2020**, *1*, 0862. [CrossRef]
44. Meyer, J.; Agathocleous, N.; Youmans, H.; Williams, H.; Vespa, T.; Rugh, J.; Lustbader, J.; Titov, E. Advanced Climate System for EV Extended Range. In *US Department of Energy Technical Report*; U.S. Department of Energy: Washington, DC, USA, 2017. [CrossRef]
45. Arasu, M.; Ahmed, Q.; Rizzoni, G. Optimizing Battery Cooling System for a Range Extended Electric Truck. *SAE Tech. Pap.* **2019**, *1*, 158. [CrossRef]
46. Kennedy, J.; Eberhart, R. Particle swarm optimization. In Proceedings of the ICNN'95—International Conference on Neural Networks, Perth, WA, Australia, 27 November–1 December 1995; Volume 4, pp. 1942–1948.
47. ANCI. Mobilità Sostenibile—Ricerca ANCI, Nelle Città Serpentoni di Auto 'Vuote', ma Cala Inquinamento. Available online: http://www.anci.it/mobilita-sostenibile-ricerca-anci-nelle-citta-serpentoni-di-auto-vuote-ma-cala-inquinamento/ (accessed on 7 June 2021).
48. Ford, D. As Cars Are Kept Longer, 200,000 Is New 100,000. *New York Times*, 16 March 2012.
49. Bloomberg. Gasoline Prices around the World: The Real Cost of Filling Up. Available online: https://www.bloomberg.com/graphics/gas-prices/#20202:Italy:EUR:l (accessed on 3 September 2021).
50. Eurostat. Electricity Prices for Household Consumers—Bi-Annual Data. Available online: https://ec.europa.eu/eurostat/databrowser/view/nrg_pc_204/default/table?lang=en (accessed on 3 September 2021).
51. Hamza, K.; Laberteaux, K.; Chu, K.-C. On Modeling the Total Cost of Ownership of Electric and Plug-in Hybrid Vehicles. *SAE Tech. Pap.* **2020**, *1*, 1435. [CrossRef]
52. Anselma, P.G.; Belingardi, G. Multi-objective optimal computer aided engineering of hydraulic brake systems for electrified road vehicles. *Veh. Syst. Dyn.* **2020**, 1–22. [CrossRef]
53. Brandimarte, P. Optimization Model Solving. In *An Introduction to Financial Markets: A Quantitative Approach*; John Wiley & Sons Inc.: Hoboken, NJ, USA, 2018; pp. 717–718.
54. Nouiri, M.; Bekrar, A.; Jemai, A.; Niar, S.; Ammari, A.C. An effective and distributed particle swarm optimization algorithm for flexible job-shop scheduling problem. *J. Intell. Manuf.* **2015**, *29*, 603–615. [CrossRef]
55. Mashadi, B.; Khadem Nahvi, M. Fuel consumption reduction by introducing best-mode controller for hybrid electric vehicles. *Proc. Inst. Mech. Eng. Part D J. Automob. Eng.* **2020**, *234*, 810–822. [CrossRef]
56. Yarpiz. Particle Swarm Optimization in MATLAB. Available online: https://yarpiz.com/50/ypea102-particle-swarm-optimization (accessed on 7 June 2021).

applied sciences

MDPI

Article

Predictive Energy Management Strategy for Hybrid Electric Air-Ground Vehicle Considering Battery Thermal Dynamics

Zhe Li [1,2], Xiaohong Jiao [1], Mingjun Zha [2], Chao Yang [2,*] and Liuquan Yang [2]

1 Electrical Engineering, Yanshan University, Qinhuangdao 066004, China
2 School of Mechanical Engineering, Beijing Institute of Technology, Beijing 100081, China
* Correspondence: cyang@bit.edu.cn

Abstract: Hybrid electric air-ground vehicles (HEAGVs), which can run on the land and fly in the air, are considered a promising future transportation. The operation of HEAGVs, accompanied by high energy consumption, could lead to increasing battery temperature, which may affect the lifespan of the battery. To make the battery last longer and improve energy efficiency, an effective energy management strategy (EMS) is necessary for the operation of HEAGVs. In this regard, this paper proposes a predictive EMS based on model predictive control (MPC). Firstly, speed information is obtained by intelligent network technology to achieve a prediction of power demand, and then the state of charge (SOC) reference trajectory is planned. Secondly, a Pontryagin's minimum principle-based model predictive control (PMP-MPC) framework is proposed, including battery thermal dynamics. Under the framework, fuel efficiency is improved by reducing the temperature of the battery. Finally, the proposed method is compared to PMP, dynamic programming (DP), and rule-based (RB) methods. The effect of different preview horizon sizes on fuel economy and battery temperature is analyzed. Verification results under two driving cycles indicate that compared with the rule-based method, the proposed method improves fuel economy by 5.14% and 5.2% and decreases the temperature by 5.9% and 4.9%, respectively. The results demonstrate that the proposed PMP-MPC method can effectively improve fuel economy and reduce temperature.

Keywords: hybrid electric air-ground vehicle; energy management strategy; battery temperature; model predictive control; Pontryagin's minimum principle

Citation: Li, Z.; Jiao, X.; Zha, M.; Yang, C.; Yang, L. Predictive Energy Management Strategy for Hybrid Electric Air-Ground Vehicle Considering Battery Thermal Dynamics. *Appl. Sci.* **2023**, *13*, 3032. https://doi.org/10.3390/app13053032

Academic Editors: Michel De Paepe and Dong-Won Kim

Received: 7 January 2023
Revised: 6 February 2023
Accepted: 24 February 2023
Published: 27 February 2023

1. Introduction

It is acknowledged that the existing traditional vehicles present challenges in contexts of traffic congestion and complex working conditions, such as bridge cutoff, mountain range, and fire rescue [1]. Air-ground vehicles (AGVs) are expected to provide a plausible solution to the above issues, which are gaining increasing attention from the aviation and automotive fields [2]. AGVs require a huge amount of energy when converting between driving and flying, especially in flight mode [3]. High fuel consumption, together with carbon emissions and environmental pollution, is one of the key weaknesses of traditional AGVs. With the development of electrical technology, AGVs are using electrical energy; however, the energy density of the on-board battery is too low and there is a range anxiety problem. The hybrid powertrain could solve the above problems to become the most suitable choice.

The past few decades have witnessed tremendous effort toward hybrid unmanned aerial vehicles and aircraft. Hybrid powertrain has received wide research attention and gradually become a research hotspot [4,5]. Hybrid powertrain can be divided into series, parallel, and series-parallel. In this case, the series configuration allows the coupling of multiple energy sources in the form of electrical energy. The series hybrid unit is electrically connected to the load. The structure of multi-rotor AGVs is characterized by symmetrical distribution around the motor propellers. By using a series architecture to design the hybrid powertrain, each motor-propeller unit can be controlled individually. The rotor and engine are not mechanically connected but electrically connected, and the space structure is relatively simple. Series hybrid electric air-ground vehicles (HEAGVs) are available as an optional solution.

Rational distribution of power between the engine-generator set and the battery can improve the fuel economy of HEAGV, which is the main task of the energy management strategy (EMS). EMSs are divided into rule-based (RB) strategies and optimization-based strategies. RB EMSs rely on personnel experience, which makes it difficult to ensure optimal control is achieved. There are two kinds of rule-based EMSs, based on the logical threshold EMSs [6] and fuzzy EMSs [7]. The EMS based on logic threshold has the advantage of simple control logic. There are two types of optimization-based control strategies: global optimization control and real-time optimization control [8]. To obtain globally optimal results, global optimization algorithms such as dynamic programming (DP) [9] and Pontryagin's minimum principle (PMP) [10] are used to solve the problem. DP can achieve the optimal global solution; however, sometimes a large amount of calculation leads to dimensional disaster. Liu et al. [11] used the DP method to determine the best torque distribution scheme for the whole driving cycle in the form of segmented decision making. The PMP method is also a global optimization method and is often considered an alternative to DP. DP can be used as a benchmark for other algorithms. However, these algorithms require complete driving cycle information in advance and are computationally intensive. These methods cannot be applied online and in real time. The equivalent consumption minimization strategy (ECMS) is a real-time optimization algorithm [12]. However, the equivalent factor is susceptible to the driving cycle, and the optimal control results take work to obtain. The model predictive control (MPC) method can be regarded as a combination of global and real-time optimization methods. Zhang et al. [13] proposed an exponentially varying speed prediction method based on MPC for power allocation. Based on deep learning methods, Wei et al. [14] developed a set success rate predictor consisting of input state classification and sub-predictors. Intelligent network technology, combined with a variety of advanced technologies, establishes connections between vehicles, drivers and road conditions, which is an effective means of solving EMS problems. Considering future traffic information, Azadi et al. [15] proposed a vehicle automatic cruise control system to reduce red light parking time and fuel consumption. Rajkumar [16] obtained the speed limit and road topography information as constraints for MPC speed prediction through intelligent network technology and calculated the safe speed as a reference. Lorenzo [17] planned the speed curve based on V2X and tested the communication between controllers and intelligent transportation systems (ITS) over cellular networks.

For a hybrid powertrain, the battery pack is a critically important part. Lower battery temperature inhibits battery discharge capacity. High battery temperature affects the battery life and has an impact on the stable operation of HEAGV. Tang et al. [18] added the battery temperature as an optimization term to the cost function, which can achieve a balance between economy and battery aging. Wang et al. [19] illustrated the enhancement mechanism of the thermal conductivity of composite phase change materials from the perspective of microstructure evolution. Ali Asef et al. [20] proposed a novel battery heat dissipation model to model and optimize battery thermal management for series hybrid electric vehicles under different standard driving cycles. Zhang et al. [21] reviewed the battery thermal management methods and compared the advantages and disadvantages of each method. Rodrigo et al. [22] proposed a new definition of the operating point for discharging Li-ion batteries based on degradation, autonomy, and heat generation to maximize battery life. Harrison and Charles [23] predicted the thermal performance and thermal runaway risk of the battery pack during electric vertical takeoff and landing operation through multi-physics field system simulations. Battery temperature is also very important for HEAGVs. Current EMSs that consider the thermal dynamics of the battery in HEAGVs are less researched. Therefore, there is still room for improvement in EMSs.

The SOC reference trajectory in hybrid electric vehicles is important for the MPC of EMSs and has been investigated by several researchers. Zhou et al. [24] aimed to divide the driving cycle into multiple segments and determine the reference SOC trajectory for each segment by a simple analytical formula. Guo et al. [25] used Q-learning to quickly plan the optimal SOC reference. He et al. [26] used DDPG to plan SOC reference trajectory adaptively and quickly. However, the driving cycles of HEAGV include air and ground phases, and the SOC reference trajectory is different from that of hybrid electric vehicles. When HEAGV is operating on the ground, the engine-generator set charges the battery, and the SOC curve tends to rise in preparation for the flight mode. When HEAGV is flying in the air, it consumes much electrical energy. The SOC changes drastically and shows a downward trend. Under the mode transition, the SOC reference trajectories in the two modes are different and segmented for planning. Therefore, it is necessary to plan the SOC reference trajectory of HEAGV.

To bridge these gaps, this paper proposes an MPC-based EMS considering battery thermal dynamics. This paper has the following contributions. First, speed information is obtained through intelligent network technology to achieve the prediction of demand power, and then the SOC reference trajectory is planned in segments. Second, a PMP-MPC EMS framework is proposed. Third, the thermal dynamics of the battery are considered within the PMP-MPC framework. Finally, the effectiveness and superiority of the proposed HEAGV control strategy are verified by comparing them with different methods.

The rest of this paper is as follows. In Section 2, the series hybrid powertrain model is established, including the battery thermal model. In Section 3, the PMP-MPC framework is described, which includes speed acquisition through intelligent network technology, and then the SOC reference trajectory is planned. Finally, the conclusion is summarized.

2. Descriptions of HEAGV with Series Hybrid Powertrain

This study was conducted on a HEAGV with a series hybrid electric propulsion system, as shown in Figure 1. HEAGV has two modes (flight mode and ground mode), which can be switched freely. The series hybrid power system comprises a turboshaft engine, battery packs, a generator, four hub motors, and sixteen rotor motors. The turboshaft engine is completely decoupled from the drive system and can run under its best working conditions to achieve high efficiency. The engine generator set and battery packs are power sources for the hub and rotor motors. Eight pairs of coaxial rotors are used to generate lift in the flight mode of the HEAGV. In ground mode, the HEAGV is propelled by hub motors like a vehicle. The parameters of HEAGV are shown in Table 1.

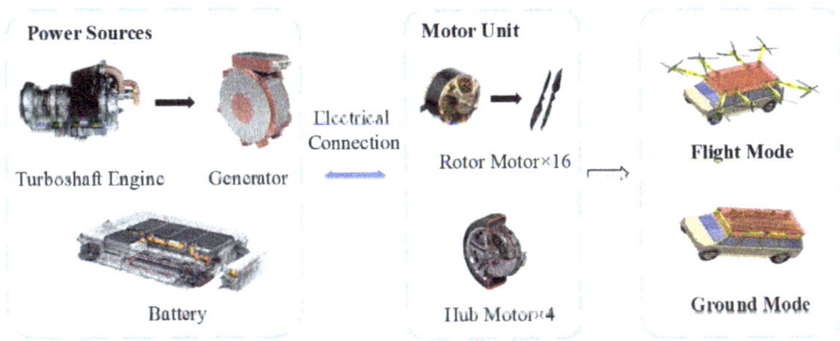

Figure 1. Structure of the studied HEAGV and powertrain system.

Table 1. Basic HEAGV parameters.

Parameter	Value	Title 3
Powertrain system	Series Hybrid Electric	Air/Ground
Vehicle mass (kg)	2000	Air/Ground
Gravity acceleration (m/s^2)	9.81	Air/Ground
Wheel radius (m)	0.39	Ground
Frontal area (m^2)	7.8	Ground
Paddle radius (m)	0.9	Air
Vertical area while climbing/landing (m^2)	32	Air
Equivalent frontal area in cruise (m^2)	12.4	Air
Engine peak power (kW)	110	Air/Ground
Rotor-motor peak power (kW)	50	Air
Hub motor peak power (kW)	60	Ground
Battery-rated capacity (Ah)	150	Air/Ground
Battery-rated voltage (V)	500	Air/Ground

2.1. Air-Ground Vehicle Model

When HEAGV is driving on the ground, according to the vehicle longitudinal dynamics, the power demand of each wheel can be calculated as follows.

$$T_w = 0.25[mgfcos\theta + 0.5AC_d\rho_{air}v^2 + mgsin\theta + \delta m\frac{dv}{dt}]r \tag{1}$$

where m denotes the air-ground vehicle mass, g refers to the gravity acceleration, f denotes the rolling resistance coefficient, θ represents the slope of the road, A stands for the fronted area, C_d shows the air drag coefficient, ρ_{air} is the air density, v is the vehicular speed, δ is the mass factor caused by the rotating inertia of wheels and powertrain rotating components, r is the wheel radius, v is the HEAGV's speed along the longitudinal direction. In addition, the hub motor, and mechanical brakes together provide the torque required by the wheels, and the powertrain torque balance can be expressed as

$$T_w = {\eta_T}^{sign(T_m)} T_m + T_b \tag{2}$$

where η_T and T_m are transmission efficiency and hub motor torque, respectively. Since there is no reduction mechanism, the transmission efficiency is set to 1.

When the HEAGV is flying in the air, according to flight aerodynamics [27], the force F_{ver} required for vertical flight is composed of the force G, the acceleration resistance F_{acc}, and the drag force F_{air}, the formula of force in vertical flight is expressed as

$$\begin{aligned} F_{ver} &= G + (F_{acc} + F_{air}) \times sign(V_{ver}) \\ &= mg + (\delta m a_v + \tfrac{1}{2}A_u C_D \rho_\alpha V_{ver}^2) \times sign(V_{ver}) \end{aligned} \tag{3}$$

$$sign(V_{ver}) = \begin{cases} 1, & \text{climb} \\ -1, & \text{descend} \end{cases} \tag{4}$$

where δ is the mass factor caused by the rotational inertia of the rotating part, a_v is the vertical acceleration speed, A_u is the vertical area, C_D is the air drag coefficient, ρ_α is the density of air, V_{ver} is the vertical climb/landing velocity.

When the HEAGV is flying horizontally in the air, the force is

$$F_{cru} = \sqrt{{F_{hor}}^2 + {F_{ver}}^2} \tag{5}$$

$$F_{hor} = \delta m a_{hor} + \frac{1}{2}A_f C_D \rho_\alpha V_{hor}^2 \tag{6}$$

2.2. Engine Model

The experimental data modeling is simple, the simulation effect is superior, and the dynamic characteristics of the turboshaft engine can be ignored. Therefore, this paper chooses an experimental modeling method to model the turboshaft engine. The engine is controlled to make it operate at a fixed speed. The turboshaft engine fuel consumption map is shown in Figure 2. The instantaneous fuel consumption of the engine is determined by the engine speed and torque as follows

$$\dot{m}_f = f(\omega_g, T_g) \tag{7}$$

The turbine drives the generator to create electrical power, and its output power P_e can be calculated as

$$P_e = \omega_e T_e \eta_e \tag{8}$$

where ω_e, T_e and η_e are turbine engine rotational speed, torque, and efficiency, respectively.

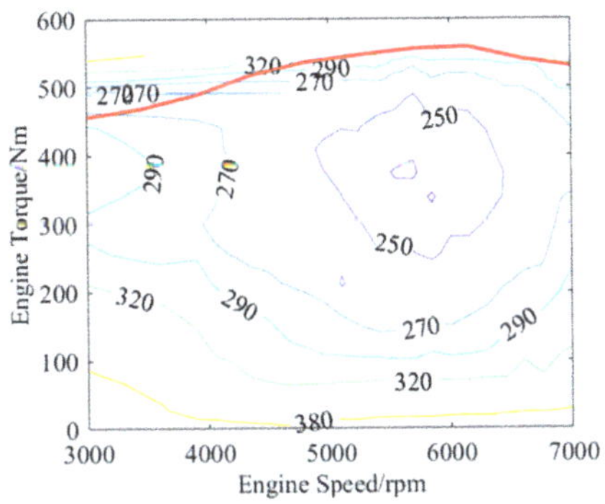

Figure 2. Fuel consumption map of the turboshaft engine.

2.3. Motor/Generator Models

In the powertrain system, the generator is used to provide electrical power. The hub motor and rotor motor drive the wheels and rotor, respectively. The motors and generators are modeled using a quasi-static model. The efficiency characteristics of the motor and generator are represented by nonlinear three-dimensional plots of torque and rotational speed using data obtained from the manufacturer. The motor efficiency plots are represented in Figures 3 and 4, respectively. The generator efficiency plots are represented in Figure 5. The motor efficiency η_m at the operation point (n_m, T_m) is calculated according to the following correlation

$$\eta_m(n_m, T_m) = f(n_m, T_m) \tag{9}$$

$$\eta_g(n_g, T_g) = f(n_g, T_g) \tag{10}$$

where η_m is the motor efficiency, η_g is the generator efficiency, n_m is the rotating speed of the motor, and n_g is the rotating speed of the generation.

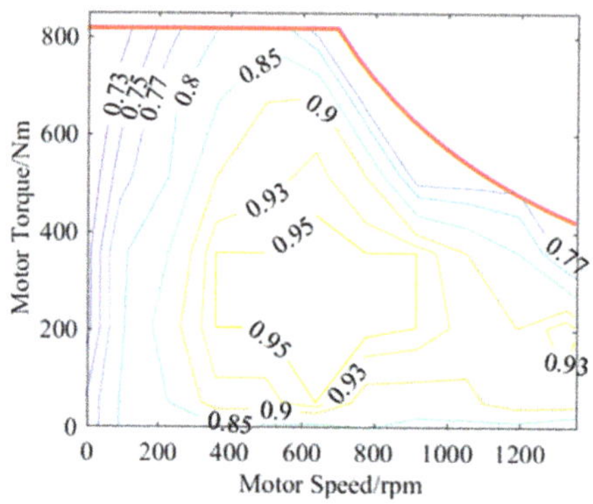

Figure 3. Wheel motor efficiency map.

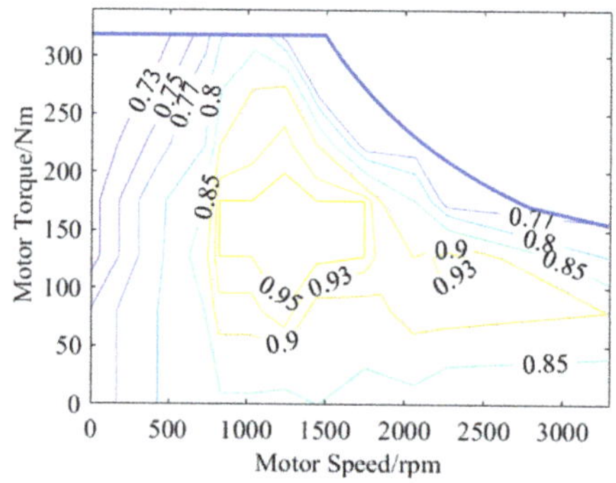

Figure 4. Rotor motor efficiency map.

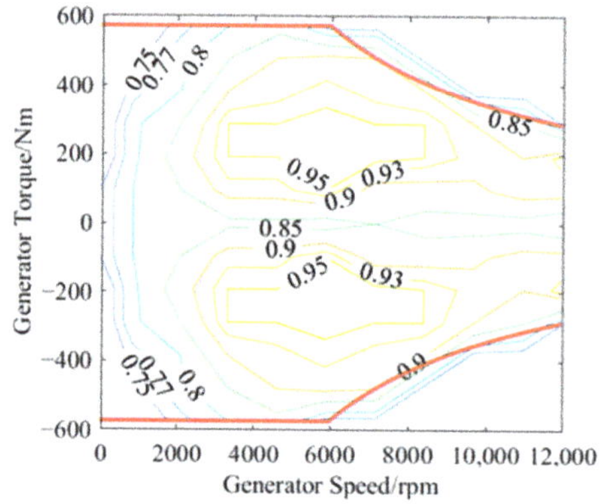

Figure 5. Generation efficiency map.

2.4. Battery Model

The battery packs are modeled as an equivalent circuit model using the opening circuit voltage U_{oc} and the internal resistance R_0, which is depicted as

$$I_{bat} = \frac{U_{oc} - \sqrt{U_{oc}{}^2 - 4R_0 P_{bat}}}{2R_0} \tag{11}$$

$$\dot{SOC} = -\frac{I_{bat}}{Q_{bat}} \tag{12}$$

where I_{bat} is the battery current, $\dot{SOC}$ is the dynamics of the battery SOC, and Q_{bat} is the battery normal capacity. Temperature is an important factor that affects battery life. To study the thermal management of the battery, the heat production of the battery itself is assumed to be evenly distributed in space. The material of the battery is uniform. The radiant heat produced by the battery is ignored. The effect of flight altitude on battery temperature is not considered. Therefore, the total heat generated by the battery Q is composed of joule heat and reaction heat, which can be expressed as [28]

$$Q = I^2 r + IT\frac{dE}{dT} \tag{13}$$

where I is the current of the battery cell, r is the internal resistance of the battery cell, T is the temperature of the battery cell, and dE/dT is the temperature influence coefficient of the battery, which is a physical quantity related to electrochemistry [29].

The heat generation rate of the battery is expressed as

$$q = \frac{Q}{V} = \rho C_p \frac{dT}{dt} \tag{14}$$

where q is the heat generation rate of the battery, ρ is the average density of the battery, C_p is the specific heat capacity of the battery, and dT/dt is the dynamics of the battery temperature.

From Equations (13) and (14), the dynamics of the battery temperature can be obtained:

$$\frac{dT}{dt} = \sigma \frac{I^2 r + IT\frac{dE}{dT}}{\rho C_p V} \tag{15}$$

where σ is the heat dissipation influence coefficient, which is related to the heat dissipation system of HEAGV batteries.

3. PMP-MPC Energy Management Strategy for HEAGV

The HEAGV needs a lot of energy to take off vertically, and the hybrid pack and battery work together to provide the power needed. The energy required for level flight is constant. Compared to flight mode, the amount of energy expended on land is relatively small. In this section, a PMP-MPC EMS framework is proposed. First, In the intelligent network environment, HEAGV establishes information interaction with cloud and ground workstations and provides HEAGV with accurate speed information through environment-aware navigation and positioning technologies. Second, in the MPC framework, SOC reference trajectories are required for each prediction horizon, and SOC reference trajectories are segmented and planned. The energy management problem for each prediction horizon can be formulated as a multi-objective multi-constraint nonlinear optimization problem. Finally, the PMP algorithm is used to allocate the power and solve the two-point boundary value problem by the dichotomous method. The overall control framework of the proposed method is shown in Figure 6.

3.1. Speed Information Acquisition

In this paper, intelligent network technology is used to obtain HEAGV speed information, as shown in Figure 6. Through V2V and V2I communication technologies, multi-dimensional connection information is obtained, including the driving status of the vehicle in front and the self-vehicle, as well as traffic signals. HEAGV can use ITS and GPS to obtain traffic information. In addition, sensors such as radar can provide the necessary status of the vehicle. The ground control station interacts with the vehicle and the cloud through wireless communication. Therefore, this effective information can be integrated into EMS to obtain accurate speed information, further improving fuel economy [30].

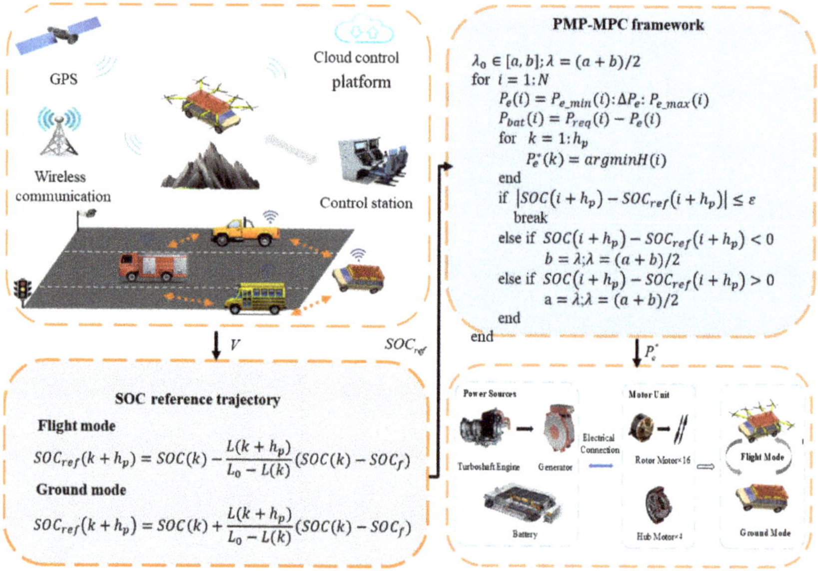

Figure 6. The overall control strategy framework.

3.2. SOC Reference Trajectory Planning

If the SOC reference trajectory is appropriate, the optimized performance will be significantly improved in the MPC framework of EMS [31]. SOC decreases roughly linearly with increasing hybrid electric vehicle driving distance. However, the driving cycle of HEAGV includes both air and ground phases. HEAGV has a higher power requirement in flight mode than in ground phases which can lead to a large amount of electrical energy being consumed. If SOC has a linear downward trend, increasing the driving distance may exceed the SOC minimum. Therefore, the SOC reference trajectory of HEAGV is different from that of hybrid vehicles. The high power requirement of HEAGV while in flight mode leads to a large amount of electrical energy being consumed. The battery is discharged, and the SOC is on a decreasing trend. The SOC tends to increase when the HEAGV is in ground mode when the battery is charged to ensure proper travel. Therefore, this paper proposes a fast planning method for global SOC reference trajectory based on traffic information. The ITS system is used to obtain the average speed information of different road sections in the driving route and to segment the different modes of HEAGV. SOC reference trajectory is divided into two types according to the HEAGV working mode. Different formulae calculate the change of SOC of different segments. The SOC reference trajectory segments of HEAGV in flight and ground modes are expressed as follows.

$$SOC_{ref}(k+h_p) = SOC(k) - \frac{L(k+h_p)}{L_0 - L(k)}(SOC(k) - SOC_f) \quad (16)$$

$$SOC_{ref}(k+h_p) = SOC(k) + \frac{L(k+h_p)}{L_0 - L(k)}(SOC(k) - SOC_f) \quad (17)$$

where k is the current time step, h_p is the size of preview horizon, v is the actual vehicular speed, v_p is the predicted speed in the preview horizon, $L(k)$ and $L(k+h_p)$ indicate the distance traveled up to $SOC(k)$, and $SOC(k+h_p)$ refers to the initial and final values SOC boundary in a preview horizon.

The SOC reference trajectory is shown in Figure 7. In the first stage, the SOC curve decreases when the HEAGV is in flight mode. In the second stage, the SOC curve rises and HEAGV is charging in ground mode to prepare for the next flight. In the third stage, HEAGV switches to flight mode, consuming a large amount of power, and the SOC curve decreases.

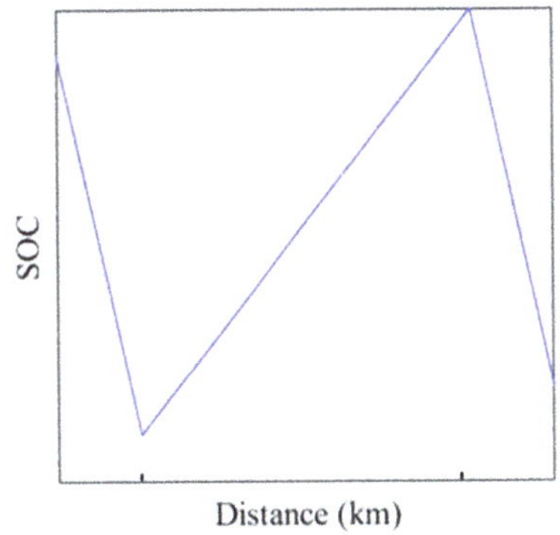

Figure 7. SOC reference trajectory.

3.3. PMP-MPC Framework

The required power of HEAGV is expressed as

$$P_{dem} = P_{eg} + P_{bat} \tag{18}$$

where P_{eg} is the power of the engine generator set, and P_{bat} is the battery power. Reasonable allocation and prediction of P_{eg} and P_{bat} is the key to achieving excellent performance.

The EMS, based on optimal control theory, seeks to minimize a global criterion over the total length of the trip. The criterion is defined by a cost function, most commonly the fuel consumption only. The cost function proposed in this paper includes the additional cost of battery temperature evolution in addition to fuel consumption. The PMP algorithm is used to solve the minimum energy consumption cost, and the objective function can be defined as

$$J = \int_{t_k}^{t_k+t_p} (J_m + J_T)dt = \int_{t_k}^{t_k+t_p} (c_f\dot{m}_f + \kappa\frac{dT}{dt})dt \tag{19}$$

where J is total cost function, J_m and J_T are fuel cost and battery temperature cost, respectively, k is current time step, t_p is the length of the prediction horizon, and c_f is the prices of fuel. dT/dt is the dynamics of the battery temperature, and κ is a weighting parameter depending on the battery temperature, which is related to battery temperature, as shown in Figure 8. The key to this additional cost is to penalize commands that cause the battery temperature to move away from its slow-aging operating range and to support commands that bring the battery temperature closer. The weighting factor κ will allow a trade-off between fuel consumption and safe battery temperature. There is no additional cost to run the battery at a safe temperature, with κ set to 0. On the other hand, when the temperature is too high, κ increases, and the higher the temperature, the higher the cost to prevent the temperature from rising further. In contrast, when the temperature is too low, the value of κ is negative, which is conducive to heating the battery and making the battery closer to the safe temperature. The optimum temperature range is 10~35 °C [32].

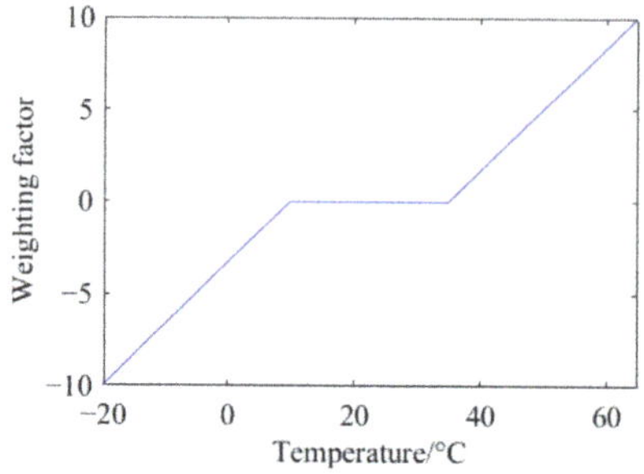

Figure 8. Influence of weighting factor κ on battery temperature.

The Hamilton function can be defined as

$$H = c_f \dot{m}_f + \kappa \frac{dT}{dt} + \lambda_1(t) S\dot{O}C + \lambda_2(t) \dot{T} \tag{20}$$

where $\lambda_1(t)$ and $\lambda_2(t)$ is the co-state variable, the state variable $x(t)$ and control variable $u(t)$ are shown as

$$x_1(t) = SOC(t) \tag{21}$$

$$x_2(t) = T(t) \tag{22}$$

$$u(t) = P_e(t) \tag{23}$$

The dynamics of the state and co-state variables can be accounted for as

$$\dot{x}(t) = \frac{\partial H}{\partial \lambda(t)} = -\frac{U_{oc} - \sqrt{U_{oc}{}^2 - 4R_0 P_{bat}}}{2R_0} \tag{24}$$

$$\dot{\lambda_1}(t) = -\frac{\partial H}{\partial SOC} = -C_f \frac{\partial \dot{m}_f}{\partial SOC} - \frac{\partial(\kappa \dot{T})}{\partial SOC} - \lambda_1(t) \frac{\lambda S\dot{O}C}{\partial SOC} - \lambda_2(t) \frac{\lambda \dot{T}}{\partial SOC} \tag{25}$$

$$\dot{\lambda_2}(t) = -\frac{\partial H}{\partial T} = -C_f \frac{\partial \dot{m}_f}{\partial T} - \frac{\partial(\kappa \dot{T})}{\partial T} - \lambda_1(t) \frac{\lambda S\dot{O}C}{\partial T} - \lambda_2(t) \frac{\lambda \dot{T}}{\partial T} \tag{26}$$

To obtain the global optimal solution, the Hamiltonian function must satisfy the following necessary constraint

$$H(\boldsymbol{x}^*(t), u^*(t), \lambda^*(t), t) \leq H(\boldsymbol{x}(t), u(t), \lambda(t), t) \tag{27}$$

Furthermore, the state variable must satisfy the following boundary conditions:

$$SOC(t_k) = SOC_{\text{int}} \tag{28}$$

$$SOC(t_k + t_p) = SOC_{end} \tag{29}$$

where SOC_{int} and SOC_{end} are the initial and final SOC values of each prediction layer, which are determined by the reference SOC.

Therefore, the optimal control sequence can be expressed as

$$P_e^* = argminH(\boldsymbol{x}(t), u(t), \lambda(t), t) \tag{30}$$

The physical constraints on state variables and control variables can be given as

$$n_{e\_\min} \leq n_e \leq n_{e\_\max} \tag{31}$$

$$T_{e\_\min} \leq T_e \leq T_{e\_\max} \tag{32}$$

$$P_{e\_\min} \leq P_e \leq P_{e\_\max} \quad (33)$$

$$I_{bat\_\min} \leq I_{bat} \leq I_{bat\_\max} \quad (34)$$

$$SOC_{\min} \leq SOC \leq SOC_{\max} \quad (35)$$

where min and max denote the upper and lower boundary values, respectively.

The shooting method is used to solve the two-point boundary problem [33]. The initial costate value is set to λ_0 and the optimal co-state value is obtained by dichotomy in the interval. In order to ensure the calculation efficiency, the maximum iteration number $j_{\max}$ is set.

4. Result and Discussion

In this section, the simulation results demonstrate the performance of the proposed strategy. First, to verify the effectiveness of the proposed control strategy, two driving cycles, which both include the flight phase and the ground phase are used as test driving cycles. Second, the proposed strategy is compared and analyzed with PMP, DP, and RB methods. Finally, the effectiveness of the proposed strategy is verified. Moreover, all the results were calculated on a laptop computer with a 1.60 GHz CPU and 16.00 GB of memory.

4.1. PMP-MPC Framework

The mission profiles of HEAGV in flight and on the ground under two driving cycles are shown in Figure 9. The first driving cycle A used in this paper consists of two flight phases and one ground phase, totaling 2360 s. To cross obstacles, the HEAGV switches to flight mode. In the first flight stage, HEAGV goes through the vertical takeoff stage, adjusts its attitude to start horizontal flight after reaching a certain altitude, and then completes vertical landing by hovering. In the second stage, the HEAGV switches to ground driving mode. At this stage, the battery needs to accumulate energy to meet the needs of the next flight, and the SOC curve generally shows an upward trend during this process. In the third stage, the HEAGV is used to cross obstacles, just as in the first stage. The second driving cycle B consists of two ground phases and one flight phase for a total of 1205 s. First HEAGV drives from the ground, then switches to flight mode, and finally switches to ground mode. The speed of HEAGV under two driving cycles A and B is shown in Figure 10.

4.2. Comparison of Different Methods

To verify the proposed method, PMP-MPC is compared in different prediction horizons of 5 s, 10 s, and 15 s, as shown in Figure 11. The results of three different prediction horizons are shown in Tables 2 and 3. The temperature and temperature change rate of the battery in the 5 s prediction horizon are shown in Figures 12 and 13. When the prediction horizon is 5 s, the total cost is the lowest and the best results are obtained. The prediction horizon is too long to affect the accuracy of the prediction. In flight mode, the engine and battery run at high power, and the battery temperature rises faster. When in ground mode the engine runs to charge the battery, and the battery temperature rises slowly. Different driving cycles under the same method have the same curve trend in both flight and ground states, which verifies the reasonableness of the method.

Table 2. Results of PMP-MPC with three prediction horizons in driving cycle A.

Horizon (s)	FC (L)	Temperature (°C)
5	78.47	29.42
10	78.94	29.07
15	79.62	28.74

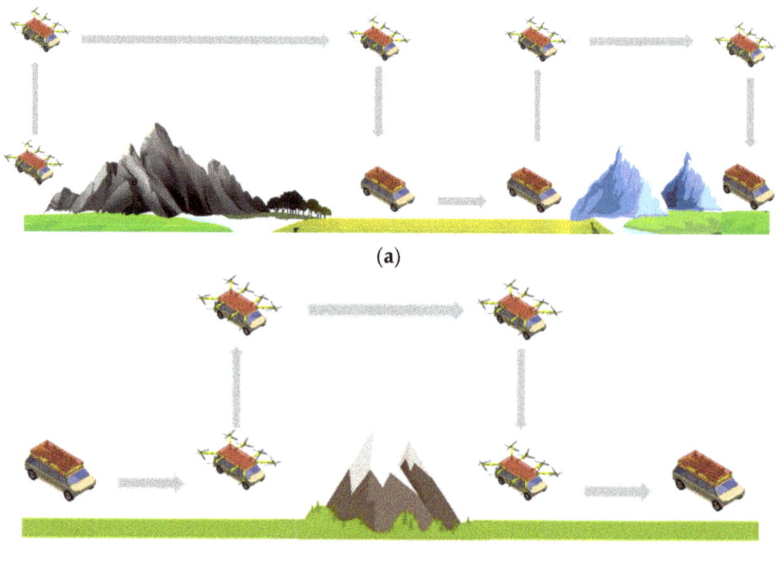

(a)

(b)

Figure 9. The HEAGV mission profile in flight and on ground. (**a**) Driving cycle A. (**b**) Driving cycle B.

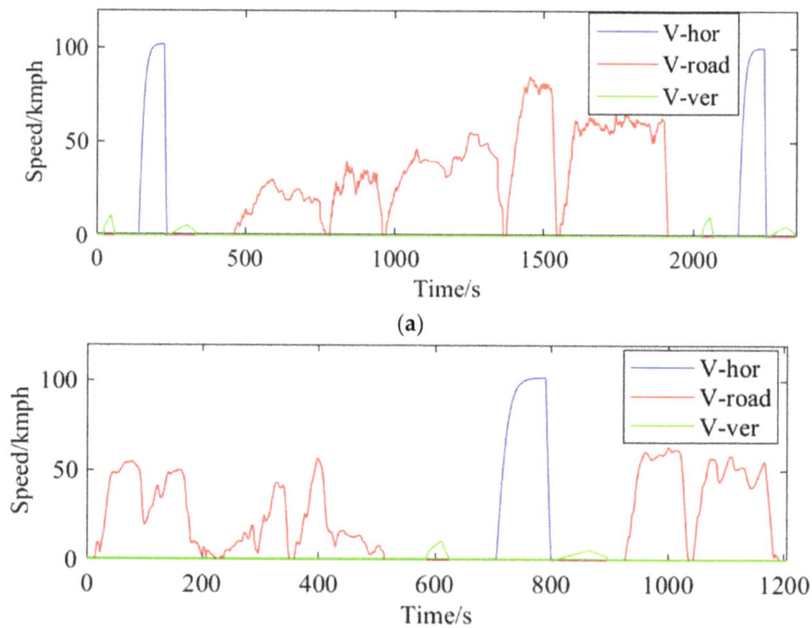

(a)

(b)

Figure 10. The HEAGV driving cycles. (**a**) Driving cycle A. (**b**) Driving cycle B. The red line represents the speed of the ground; the green line represents the vertical takeoff and landing speed; the blue line represents the speed of the horizontal cruise.

In order to further verify the proposed method, PMP-MPC, DP, PMP, and RB methods are used for comparison. The initial temperature was set at 20 °C. The comparison SOC curve is shown in Figure 14. It can be seen from the figure that SOC curves obtained by these different methods have the same general trend. The SOC curve drops in flight mode and rises in ground mode. The results show that the proposed PMP-MPC method can achieve nearly global optimal results and has the potential of online real-time application. There is a performance improvement compared to the rule-based method. The fuel consumption and battery temperature for the different methods in driving cycles A and B are shown in Figures 15 and 16. Temperature comparison only considers the real-time online method.

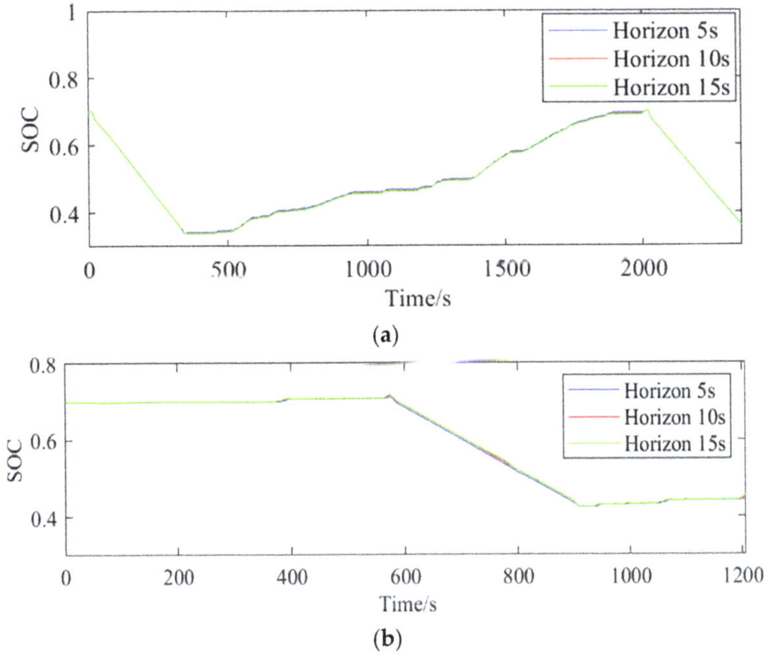

Figure 11. SOC profiles with different prediction horizon lengths. (**a**) SOC profiles with different predicted horizon lengths under driving cycle A. (**b**) SOC profiles with different predicted horizon lengths under driving cycle B.

Table 3. Results of PMP-MPC with three prediction horizons in driving cycle B.

Horizon (s)	FC (L)	Temperature (°C)
5	24.79	22.12
10	25.12	22.27
15	25.63	23.09

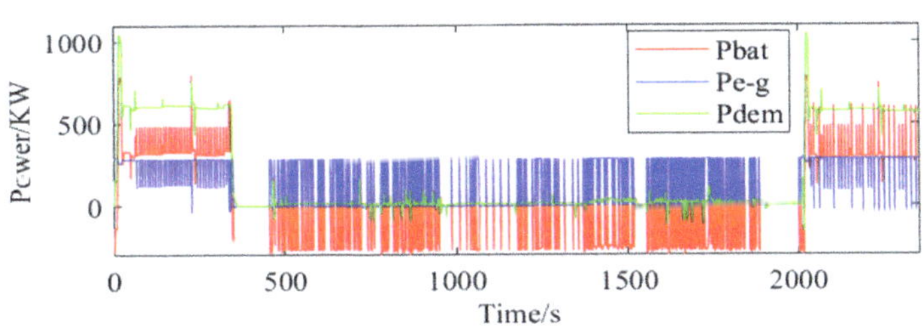

(**a**)

Figure 12. *Cont.*

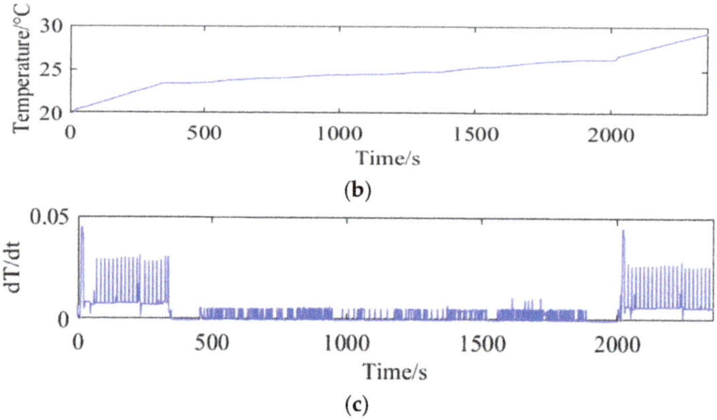

Figure 12. Validation of the PMP-MPC strategy under the air-ground driving cycle A. (**a**) The blue line represents the engine-generator set power; the red line represents the battery power; the green line represents the demand power. (**b**) Battery temperature profiles. (**c**) Battery temperature change rate per instant.

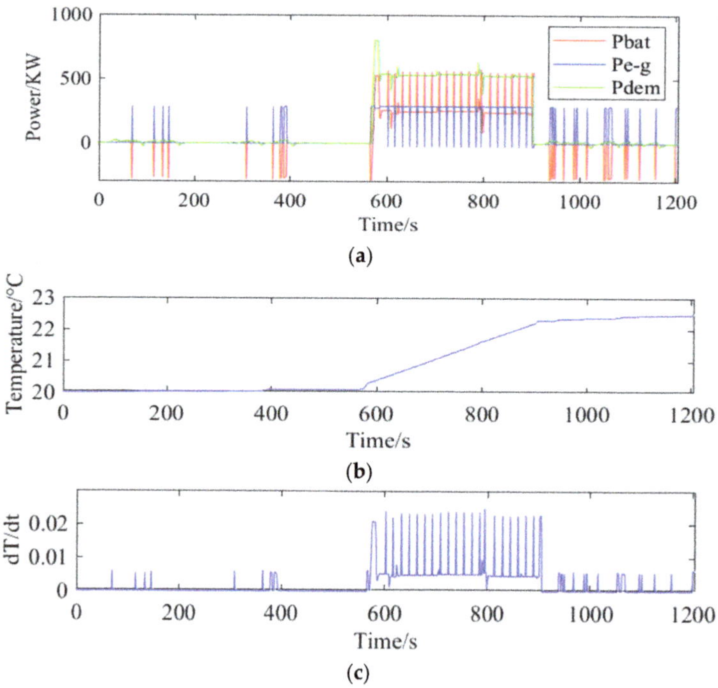

Figure 13. Validation of the PMP-MPC strategy under the air-ground driving cycle B. (**a**) The blue line represents the engine-generator set power; the red line represents the battery power; the green line represents the demand power. (**b**) Battery temperature profiles. (**c**) Battery temperature change rate per instant.

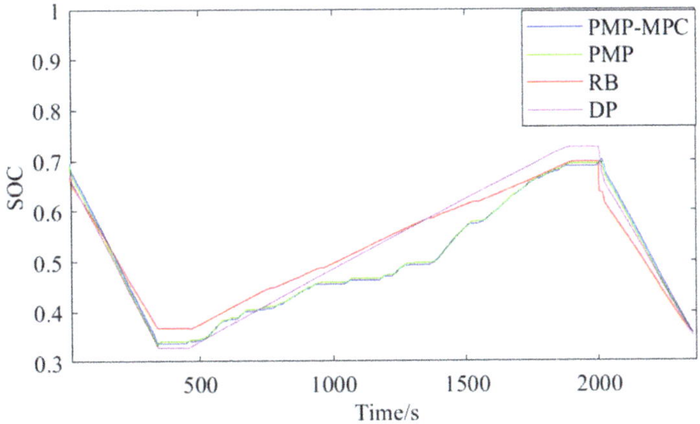

(a)

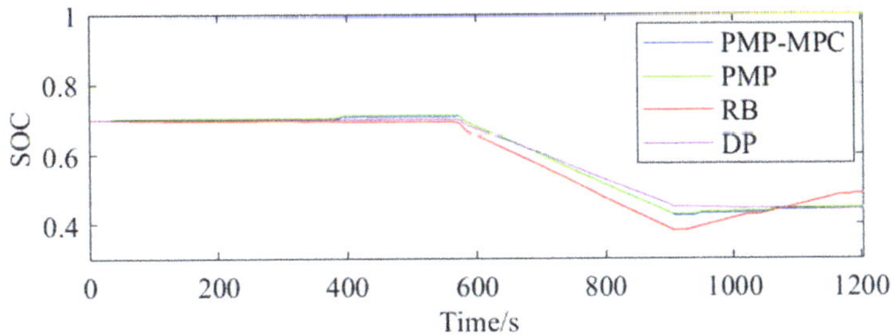

(b)

Figure 14. SOC profiles of different approaches. (**a**) SOC profiles of different approaches in driving cycle A. (**b**) SOC profiles of different approaches in driving cycle B. (Prediction horizon = 5 s for PMP-MPC).

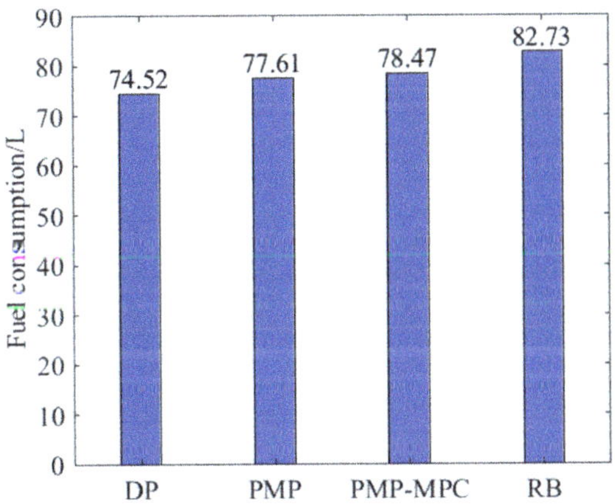

(a)

Figure 15. *Cont.*

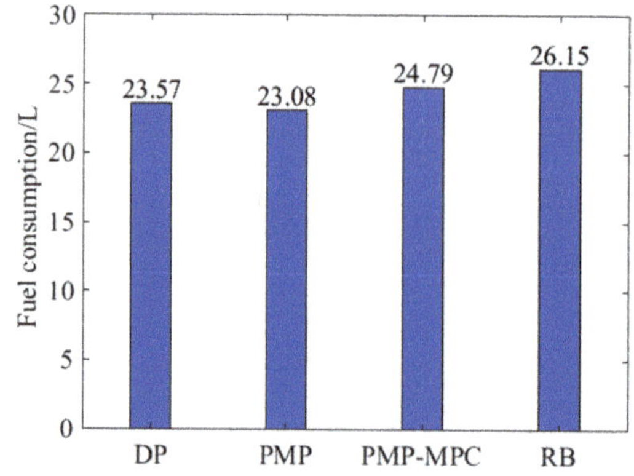

(b)

Figure 15. Fuel consumption for different methods. (**a**) Fuel consumption for different methods in driving cycle A. (**b**) Fuel consumption for different methods in driving cycle B.

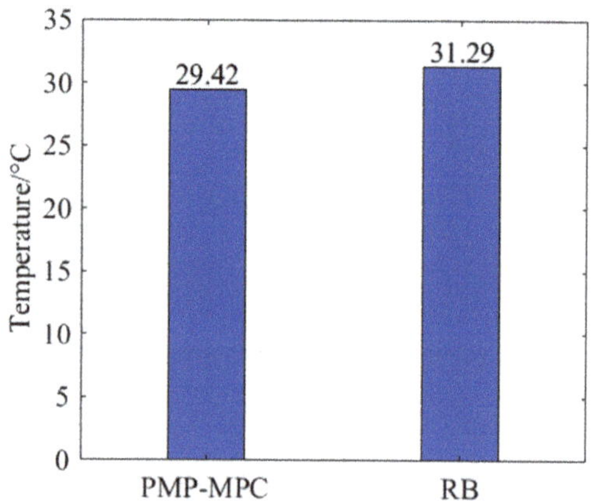

(a)

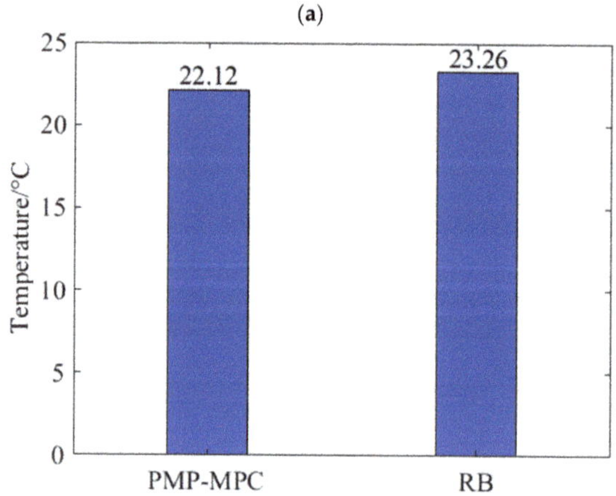

(b)

Figure 16. Battery temperatures for different methods. (**a**) Battery temperatures for different methods in driving cycle A. (**b**) Battery temperatures for different methods in driving cycle B.

The results show that the proposed approach has higher fuel consumption than the global optimization approach because MPC-based EMS is essentially a local optimization algorithm. Overall, the PMP-MPC algorithm may provide a better alternative for developing an efficient and low-cost EMS. In addition, the RB-based method, although capable of online implementation and fast computation, has a significantly higher fuel cost.

5. Conclusions

In this paper, a PMP-MPC method is proposed considering the battery thermal dynamics for HEAGVs. First, speed information is obtained through intelligent network technology to realize the prediction of power demand, and then the SOC reference trajectory is planned in segments. Second, a PMP-MPC EMS is proposed. Third, the effect of battery temperature on total cost is analyzed by considering battery temperature in the PMP-MPC EMS. The resulting EMS optimizes fuel consumption and battery temperature, and the strategy is applicable to the HEAGV. Finally, the proposed method is compared with PMP, DP, and RB to verify its effectiveness. Compared with the rule-based method, the fuel economy improved by 5.14% and 5.2% for both driving cycles a and b, respectively. Temperature costs are also reduced by 5.9% and 4.9%, respectively. Therefore, the proposed PMP-MPC is an effective online method.

Author Contributions: Conceptualization, Z.L., X.J. and C.Y.; methodology, Z.L., X.J. and M.Z.; software, Z.L. and M.Z.; validation, Z.L., X.J. and M.Z.; formal analysis, Z.L., X.J. and M.Z.; investigation, Z.L.; resources, Z.L.; data curation, Z.L.; writing—original draft preparation, Z.L., X.J. and M.Z.; writing—review and editing, Z.L., X.J., M.Z., C.Y. and L.Y.; visualization, Z.L.; supervision, Z.L. and X.J. All authors have read and agreed to the published version of the manuscript.

Funding: This work was supported by the National Natural Science Foundation of China (No. 61973265).

Institutional Review Board Statement: Not applicable.

Informed Consent Statement: Not applicable.

Data Availability Statement: Not applicable.

Conflicts of Interest: The authors declare no conflict of interest.

References

1. Jordi, P.; Tanja, Z.; Jovana, K. On the understanding of the current status of urban air mobility development and its future prospects: Commuting in a flying vehicle as a new paradigm. *Transp. Res. Part E* **2022**, *166*, 102868.
2. Qie, T.; Wang, W.; Yang, C.; Li, Y.; Liu, W.; Xiang, C. A path planning algorithm for autonomous flying vehicles in cross-country environments with a novel TF-RRT∗ method. *Green Energy Intell. Transp.* **2022**, *1*, 100026. [CrossRef]
3. Zhao, J.; Yang, C.; Wang, W.; Xu, B.; Li, Y.; Yang, L.; Zhu, H.; Xiang, C. A game-learning-based smooth path planning strategy for intelligent air-ground vehicle considering mode switching. *IEEE Trans. Transp. Electrif.* **2022**, *8*, 3142150. [CrossRef]
4. Xie, Y.; He, S.; Savvaris, A.; Tsourdos, A.; Zhang, D.; Xie, A. Convexification in energy optimization of a hybrid electric propulsion system for aerial vehicles. *Aerosp. Sci. Technol.* **2022**, *123*, 107509. [CrossRef]
5. Wang, W.; Chen, Y.; Yang, C.; Li, Y.; Xu, B.; Xiang, C. An enhanced hypotrochoid spiral optimization algorithm based intertwined optimal sizing and control strategy of a hybrid electric air-ground vehicle. *Energy* **2022**, *257*, 124749. [CrossRef]
6. Huang, K.; Nguyen, M.; Chen, P. A Rule-Based Control Strategy of Driver Demand to Enhance Energy Efficiency of Hybrid Electric Vehicles. *Appl. Sci.* **2022**, *12*, 8507. [CrossRef]
7. Yang, C.; Lu, Z.; Wang, W.; Li, Y.; Chen, Y.; Xu, B. Energy management of hybrid electric propulsion system: Recent progress and a flying car perspective under three-dimensional transportation networks. *Green Energy Intell. Transp.* **2023**, *2*, 100061. [CrossRef]
8. Liu, K.; Jiao, X.; Yang, C.; Wang, W.; Xiang, C.; Wang, W. Event-Triggered intelligent energy management strategy for plug-in hybrid electric buses based on vehicle cloud optimization. *Intell. Transp. Syst.* **2020**, *14*, 1153–1162. [CrossRef]
9. Tang, W.; Wang, Y.; Jiao, X.; Ren, L. Hierarchical energy management strategy based on adaptive dynamic programming for hybrid electric vehicles in car-following scenarios. *Energy* **2023**, *265*, 126264. [CrossRef]
10. Zhang, S.; Hu, X.; Xie, S.; Song, Z.; Hu, L.; Hou, C. Adaptively coordinated optimization of battery aging and energy management in plug-in hybrid electric buses. *Appl. Energy* **2019**, *256*, 113891. [CrossRef]
11. Liu, J.; Chen, Y.; Zhan, J.; Shang, F. Heuristic Dynamic Programming Based Online Energy Management Strategy for Plug-In Hybrid Electric Vehicles. *IEEE Trans. Veh. Technol.* **2019**, *68*, 4479–4493. [CrossRef]

12. Yang, C.; Du, S.; Li, L.; You, S.; Yang, Y.; Zhao, Y. Adaptive real-time optimal energy management strategy based on equivalent factors optimization for plug-in hybrid electric vehicle. *Appl. Energy* **2017**, *203*, 883–896. [CrossRef]
13. Zhang, S.; Xiong, R.; Sun, F. Model predictive control for power management in a plug-in hybrid electric vehicle with a hybrid energy storage system. *Appl. Energy* **2017**, *185*, 1645–1662. [CrossRef]
14. Wei, Z.; Ma, Y.; Xiang, C.; Liu, D. Power Prediction-Based Model Predictive Control for Energy Management in Land and Air Vehicle with Turboshaft Engine. *Complexity* **2021**, *2021*, 2953241. [CrossRef]
15. Asadi, B.; Vahidi, A. Predictive Cruise Control: Utilizing Upcoming Traffic Signal Information for Improving Fuel Economy and Reducing Trip Time. *IEEE Trans. Control. Syst. Technol. A Publ. IEEE Control. Syst. Soc.* **2011**, *19*, 707–714. [CrossRef]
16. Deshpande, S.R.; Jung, D.; Bauer, L.; Canova, M. Integrated approximate dynamic programming and equivalent consumption minimization strategy for eco-driving in a connected and automated vehicle. *IEEE Trans. Technol.* **2021**, *70*, 11204–11215. [CrossRef]
17. Brunelli, L.; Capancioni, A.; Gonnella, P.; Casadio, R.; Brusa, A.; Cavina, N.; Caggiano, M. A hybrid vehicle hardware-in-the-loop system with integrated connectivity for ehorizon functions validation. *IEEE Trans. Veh. Technol.* **2021**, *70*, 4340–4352. [CrossRef]
18. Tang, X.; Jia, T.; Hu, X.; Huang, Y.; Deng, Z.; Pu, H. Naturalistic Data-Driven Predictive Energy Management for Plug-in Hybrid Electric Vehicles. *IEEE Trans. Transp. Electrif.* **2020**, *7*, 497–508. [CrossRef]
19. Wang, X.; Li, B.; Qu, Z.; Zhang, J.; Jin, Z. Effects of graphite microstructure evolution on the anisotropic thermal conductivity of expanded graphite/paraffin phase change materials and their thermal energy storage performance. *Int. J. Heat Mass Transf.* **2020**, *155*, 119853. [CrossRef]
20. Asef, A.; Chitsaz, I.; Madani, N. Modeling and total cost optimization of battery thermal management system in a hybrid electric vehicle. *J. Energy Storage* **2022**, *172*, 104844. [CrossRef]
21. Zhang, X.; Li, Z.; Luo, L.; Fan, Y.; Du, Z. A review on thermal management of lithium-ion batteries for electric vehicles. *Energy* **2022**, *238*, 121652. [CrossRef]
22. Pires, R.A.; Carvalho, S.A.; Cardoso Filho, B.J.; Pires, I.A.; Huebner, R.; Maia, T.A. The Assessment of Electric Vehicle Storage Lifetime Using Battery Thermal Management System. *Batteries* **2023**, *9*, 10. [CrossRef]
23. Harrison, J.; Charles, D.; Zenker, J.; Frank, E. Using multi-physics system simulation to predict battery pack thermal performance and risk of Thermal runaway during eVTOL aircraft operations. In Proceedings of the 2019 AIAA/IEEE Electric Aircraft Technologies Symposium (EATS), Indianapolis, IN, USA, 22–24 August 2019.
24. Zhou, W.; Chen, Y.; Zhai, H.; Zhang, W. Predictive Energy Management for a Plug-in Hybrid Electric Vehicle Using Driving Profile Segmentation and Energy-based Analytical SoC Planning. *Energy* **2020**, *220*, 119700. [CrossRef]
25. Guo, L.; Zhang, X.; Zou, Y.; Guo, N.; Li, J.; Du, G. Cost-optimal energy management strategy for plug-in hybrid electric vehicles with variable horizon speed prediction and adaptive state-of-charge reference. *Energy* **2021**, *232*, 120993. [CrossRef]
26. He, H.; Huang, R.; Meng, X.; Zhao, X.; Wang, Y.; Li, M. A novel hierarchical predictive energy management strategy for plug-in hybrid electric bus combined with deep deterministic policy gradient. *Energy Storage* **2022**, *52*, 104787. [CrossRef]
27. Wang, W.; Chen, Y.; Yang, C.; Li, Y.; Xu, B.; Huang, K.; Xiang, C. An efficient optimal sizing strategy for a hybrid electric air-ground vehicle using adaptive spiral optimization algorithm. *Power Sources* **2022**, *517*, 230704. [CrossRef]
28. Bernardi, D.M.; Pawlikowski, E.M.; Newman, J. A general energy balance for battery systems. *J. Electrochem. Soc.* **1985**, *132*, 5–12. [CrossRef]
29. Onda, K.; Ohshima, T.; Nakayama, M.; Fukuda, K.; Araki, T. Thermal behavior of small lithium-ion battery during rapid charge and discharge cycles. *J. Power Sources* **2006**, *158*, 535–542. [CrossRef]
30. Wang, W.; Guo, X.; Yang, C.; Zhang, Y.; Zhao, Y.; Huang, D.; Xiang, C. A multi-objective optimization energy management strategy for power split HEV based on velocity prediction. *Energy* **2021**, *238*, 121714. [CrossRef]
31. Zhou, Q.; Du, C. A two-term energy management strategy of hybrid electric vehicles for power distribution and gear selection with intelligent state-of-charge reference. *J. Energy Storage* **2021**, *42*, 103054. [CrossRef]
32. Padovani, T.M.; Debert, M.; Colin, G.; Chamaillard, Y. Optimal energy management strategy including battery health through thermal management for hybrid vehicles. *IFAC Proc. Vol.* **2013**, *46*, 384–389. [CrossRef]
33. Ritter, A.; Widmer, F.; Duhr, P.; Onder, C.H. Long-term stochastic model predictive control for the energy management of hybrid electric vehicles using Pontryagin's minimum principle and scenario-based optimization. *Appl. Energy* **2022**, *322*, 119192. [CrossRef]

applied sciences

MDPI

Article

Numerical Performance Investigation of a Hybrid eCVT Specialized Agricultural Tractor

Francesco Mocera *,† and Valerio Martini †

Department of Mechanical and Aerospace Engineering, Politecnico di Torino, Corso Duca degli Abruzzi 24, 10129 Torino, Italy; valerio.martini@studenti.polito.it
* Correspondence: francesco.mocera@polito.it; Tel.: +39-011-0905177
† These authors contributed equally to this work.

Abstract: The need for highly efficient agricultural machineries is increasing the interest of the research community and of industrial manufacturers towards the use of integrated electric systems in combination with traditional powertrain elements. In this work, a hybrid electric tractor with electric continuously variable transmission (eCVT) capabilities was studied to investigate their performance in comparison with that of traditional diesel-powered tractor designs. This hybrid electric configuration can be classified as a power-split architecture that aims to combine the best characteristics of both the simpler parallel and the series hybrid layout while minimizing their main drawbacks. An eCVT configuration can allow for optimizing the diesel operating point with respect to the current working conditions, and achieving peak power performance and energy saving with relatively small electric machines. The proposed hybrid eCVT (HeCVT) tractor architecture was studied using a numerical model that allowed for developing two different control strategies: a charge depleting mode enabling the driver to use full power for the most power-intensive scenarios and a charge sustaining mode developed to optimize efficiency and battery use along an entire work day. To test the proposed architecture, several tasks derived from experimental field measurements on a specialized agricultural tractor were used. HeCVT results were compared with a numerical model of the traditional tractor validated by these experimental data. The HeCVT tractor showed good performance in terms of peak power capabilities using a downsized diesel engine, and consistent fuel savings were obtained according to typical daily working scenarios.

Keywords: agricultural machinery; electric vehicles; hybrid electric tractors; mechanical power transmission; NRMM

Citation: Mocera, F.; Martini, V. Numerical Performance Investigation of a Hybrid eCVT Specialized Agricultural Tractor. *Appl. Sci.* **2022**, *12*, 2438. https://doi.org/10.3390/app12052438

Academic Editors: Vincenzo Di Dio, Antonella Castellano and Marco Cammalleri

Received: 27 January 2022
Accepted: 24 February 2022
Published: 25 February 2022

1. Introduction

The growing need for food to sustain population growth requires a high level of productivity from the agricultural field, which is responsible for raw material supply. To maintain the rhythm with the entire chain, agricultural mechanization plays a crucial role in increasing farm productivity [1]. However, everything comes with a price. It is estimated that only the agricultural field is responsible for almost 20% of the annual global production of CO_2 emissions [2]. In particular, self-propelled agricultural machineries are responsible for the production of CO_2, CO, NO_x, particulate matter (PM), and hydrocarbons (HC) as side products deriving from fuel combustion within their diesel engines [3,4]. These machines are characterized by medium–high power demand and may need to run for 8–12 h per day in certain periods of the year to meet harvesting schedules. To mitigate their impact on pollutant emissions, strict regulations are progressively introduced [5] providing specifications for diesel engines emissions for all non-road mobile machineries (NRMMs) as a function of the power rating of their engines. OEM diesel engine manufacturers have focused their efforts on the development of exhaust gas aftertreatment systems [4] as a solution to comply with emissions regulations. As a matter of fact, not being able to meet those limit values would prevent their customers from sell their machines. However, those systems require increasingly more dedicated space with the increasing level of restrictions. Aftertreatment systems both do not solve the emissions problem due to side products they produce, and have introduced an increasing level of complexity when it comes to vehicle integration of such devices. Agricultural, in particular specialized, tractors, are well-known for the compactness of their powertrain, which helps in improving stability, visibility, and manoeuvrability. This is the reason why researchers and manufacturers involved in the agricultural field are investing time and money to explore new technologies to reduce emissions production, improving the overall machine efficiency for a given working task.

Thanks to the achievements in the transportation field, electrification is now one of the most promising strategy to improve machine efficiency in agricultural applications [6–8]. Electrification of an agricultural powertrain can take place at different levels depending on the specific use. Somà et al. [9] proposed a Hybridization Factor (HF) for NRMMs with the aim of defining the correct weight between the thermal engine and the electric machines in the powertrain. The authors suggest that the HF should derive from the knowledge of the work cycles the machines are expected to accomplish [10]. Although several attempts from different associations and companies [11], today the lack of standard work cycles forces researchers and manufacturers to the adoption of an experience-based design approach related to statistical features of historical and experimental data [12]. For some machines, a good understanding of the daily work cycle can bring to full electric (FE) architectures [13] where the internal combustion engine can be totally replaced by a battery-pack-based storage system, achieving an HF = 1. This, with the use of green technology for clean electricity [14], would achieve zero emissions at least in the use phase of agricultural machines' life cycle. However, in most cases, knowledge of the work cycle is not so straightforward because of the multi-purpose nature of agricultural machinery. This is very typical for agricultural tractors, which can be considered to be mobile propulsion systems to propel externally attached implements and tools [15]. Full electric tractors (FETs) are explored in the literature for small-medium sized machines. Liu et al. [16] proposed an FET powered by a low-voltage battery pack for small-farm low-power applications. Li et al. [17] proposed optimizing a dual-motor FET architecture to improve overall efficiency. In this case, the authors also considered a small–medium sized power tractor as reference for their analysis. The main reason was related to the objective difficulty in the design of a proper battery pack in heavy work-vehicle applications in terms of the trade-off in performance [18] and costs [19].

As shown in [6,20], hybrid electric tractors (HETs) represent the most feasible solution for medium-high size machines in the upcoming decade. Dalboni et al. [21] proposed an HET with parallel architecture powered by a downsized diesel engine and a 25 kWh battery pack. The authors numerically investigated the architecture against several field working scenarios with a charge-sustaining energy management strategy showing promising fuel savings. A similar parallel architecture was investigated by Mocera et al. [22,23], testing the HET model against test cases derived from field measurements on a traditional diesel-powered orchard tractor. The authors demonstrated that the proposed architecture was able to perform as well as the traditional tractor in terms of peak performance despite the downsized diesel engine, and to achieve 18% fuel saving in daily working cycles on average thanks to the proposed load observer strategy. Baek et al. [24] proposed a HET with series configuration with four dedicated wheel motors receiving power from both the battery pack and the four gasoline engine-based commercial generators. However, the range extender configuration of the architecture, a limited range of 2.5 h, was estimated for heavy plough tillage and 6 h for low-speed drive. Jia et al. [25,26] proposed a numerical investigation to optimize the use of a series architecture for agricultural tractors. Using dynamic programming to approach optimal control, the authors were able to improve fuel efficiency by about 5% with respect to results of the same architecture managed with rule-based strategies. Parallel and series configurations have advantages and disadvantages. Parallel HETs are the most promising configurations because of their relatively easy implementation on existing vehicles. However, the main limitation of parallel configuration is the mechanical coupling between ICE and mechanical load (wheels and power take off (PTO)), which prevents it from achieving the best results in terms of efficiency and emissions optimization. These problems would be solved in a series HET configuration that would instead require large electric machines to have full power on both PTO and drivetrain. This is the reason why the combined power-split configuration was recently investigated in the literature for agricultural tractors. Rossi et al. [27] explored a hybrid eCVT (HeCVT)-based tractor architecture for performance evaluation. The proposed layout consisted of an eCVT transmission inspired by the Toyota Prius layout for the driveline [28] and a direct connection of the sun gear/ICE to the PTO allowing for direct power path from ICE to attached implement. The authors defined three different operating modes for this powertrain architecture: full hybrid mode to approach full speed and heavy load; parallel hybrid mode for heavy load at low-medium speed; and EV mode for low-power application. Despite the higher complexity of the architecture and its control algorithm, the power-split configuration allowed for the engine to work at the optimal rotational speed for a given instantaneous load but at the same time gave the possibility of a joint cooperation in parallel with the electric system when needed. This is crucial to keep electric motors small and in compliance with space requirements for agricultural tractors. Tebaldi et al. [29] proposed investigating a similar eCVT powertrain applied on an agricultural tool, demonstrating ICE efficiency improvements thanks to the degree of freedom introduced with the use of the power-split configuration.

In this work, an alternative hybrid eCVT (HeCVT) configuration, such as the one proposed in [30], was investigated, and its performance was compared to that of a traditional specialized orchard tractor. The proposed numerical model was used to test the architectural characteristics according to several testing scenarios proposed by the authors in previous works [22,23], derived from experimental field tests. Two energy management strategies were defined to meet the need for daily working scenarios [31]. Thanks to the energy management strategy proposed by the authors in this paper, the HeCVT architecture for specialized orchard tractor under investigation was able to achieve the following results:

- Peak power performance comparable with those of a conventional tractor but with a downsized diesel engine and a simplified gearbox. The last point is not something that was considered in other works that wanted to keep the same tractor gearbox.
- Increased overall efficiency with fuel saving ranging from 10% in the most heavy PTO loading scenarios up to 20% in heavy trailer transportation.
- Capability for full electric operations decoupling the ICE and using the two electric machines in a dual-input-motor configuration using the energy stored in the battery pack.

2. Case Study

In this work, analysis focused on an a specialized orchard tractor, a category where compactness is crucial. These vehicles must be able to work nimbly among narrow rows and to have good curving capabilities to increase productivity. Table 1 shows the relevant characteristics of the conventional diesel-powered orchard tractor considered for this analysis.

Traditionally, these working vehicles adopt high-power diesel units to cover the widest possible range of applied loads. However, considering the most common working scenarios for these tractors, their powertrains usually work at medium–low loads far from the optimal efficiency condition, leading to unnecessary fuel consumption. In order to simulate the behavior of the conventional tractor, its main subsystems were numerically modelled with MATLAB/Simulink using a modular and multiphysics power-based approach described in previous works [22]. The main subsystems covered in this work were:

- vehicle and trailer dynamics;
- engine;
- gearbox and clutches;

- PTO loads.

Table 1. Main orchard-tractor characteristics.

Mass	2570 kg
Vehicle wheelbase	1900 mm
Track width	1850 mm
Rolling radius	680 mm
Nominal power	73 kW @ 2400 rpm
Max engine rotational speed	2400 rpm
Max vehicle speed	40 km/h
Minimal speed	0.4 km/h

The longitudinal dynamic behaviour of the tractor was modelled using a 1D model as shown in Figure 1.

$$m\dot{V}_x = 2\left(F_{xf} + F_{xr}\right) - F_d - mg \cdot sin\beta \tag{1}$$

$$F_{zf} = \frac{-h(F_d + mg \cdot sin\beta) + b \cdot mg \cdot cos\beta}{2(a+b)} \tag{2}$$

$$F_{zr} = \frac{+h(F_d + mg \cdot sin\beta) + a \cdot mg \cdot cos\beta}{2(a+b)} \tag{3}$$

where:

- a, b, and h represent the relative position of the centre of gravity of the vehicle with respect to the front and rear axles;
- m is the tractor mass; g the acceleration of gravity;
- β is the road slope angle;
- V_x is the vehicle longitudinal speed;
- F_d is the aerodynamic drag force as
  $F_d = 0.5\rho C_d A {V_x}^2 \cdot sign(V_x)$,
  ρ, air density; and C_d, drag coefficient, a frontal cross-sectional area of the vehicle. However, this contribution is almost negligible for this specific application.
- F_{xf} and F_{xr} are contact forces between wheels and ground on the longitudinal direction (front and rear axle).
- F_{zf} and F_{zr} are normal contact forces between wheels and ground (front and rear axle).

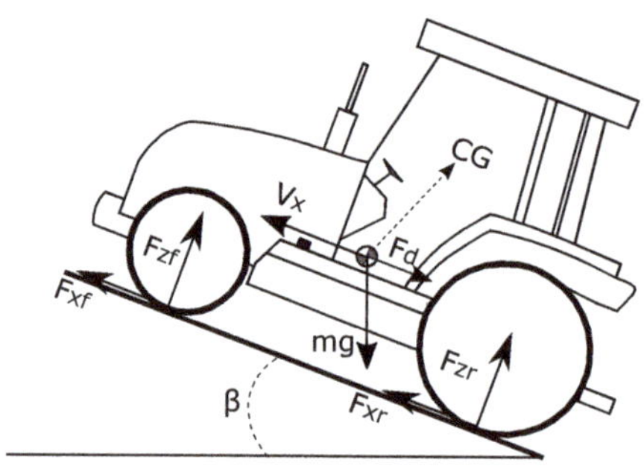

Figure 1. 1D longitudinal tractor model.

When modelling the presence of a trailer attached to the tractor, the connection between them was considered to be rigid. As a consequence, they shared the same longitudinal speed. During simulations, different payloads were considered for the trailer by changing its body mass. The engine's power output, as a combination of torque and speed, was numerically modelled using a tabulated torque data approach. Figure 2 shows the torque and power curves of the thermal unit considered in this work for the conventional powertrain. The proposed torque and power curves were derived from data available online for commercial diesel-engine units having nominal characteristics as close as possible to the ones of the reference conventional tractor studied in this and previous works by the authors.

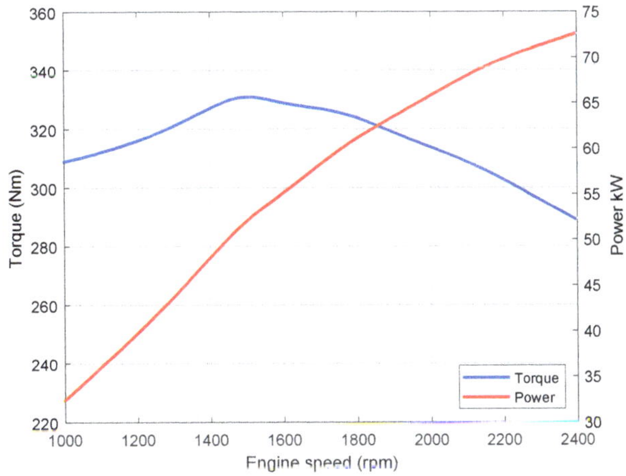

Figure 2. Torque and power characteristics for the diesel engine of a conventional tractor.

In order to compare the traditional powertrain and the proposed hybrid architecture, the engine subsystem included a multiparameter fuel consumption model that could consider both the ICE operating point characteristics and its nominal power rating. This model was developed by the Nebraska Tractor Test Laboratory and proposed by Grisso et al. [32]. This model was derived from fuel consumption measurements by the previously mentioned laboratory on more than 500 diesel engines designed for tractors. Although it gives an approximation for fuel consumption, it was considered to be sufficient since the main purpose of this work is not to evaluate the real fuel consumptions, but to make a comparison between the fuel consumptions of the conventional powertrain and the proposed hybrid architecture when facing the same tasks. This model was chosen since it correlated the rated power of the engine and its fuel consumption. The model consisted of the following equations:

$$Q_F = (0.0434 \cdot X + 0.019) \cdot P_{nom} \tag{4}$$

$$Q_R = Q_F \cdot [1 - (N - 1) \cdot (0.45 \cdot X + 0.877)] \tag{5}$$

where:

- Q_F is diesel fuel consumption at partial load and full throttle (gal/h);
- Q_R is diesel fuel consumption at partial load and reduced throttle (gal/h);
- X is the actual engine load expressed as the ratio of the actual engine output power over the maximal power available at the same speed (-);
- P_{nom} is the nominal engine power (hp);
- N is the ratio between actual engine speed and nominal speed at which P_{nom} is evaluated.

In the gearbox, tractors usually have a high number of transmission ratios. The conventional powertrain considered in this work had a total number of 32 possible gear ratios (Figure 3): two speed ranges (low–high) with four gears, a gear pair for further reduction of all previous combinations (creeper gear), a reverse gear to allow for the use of all 16 gear ratios in backward and forward motion. This complex gearbox architecture is not uncommon in agricultural tractors and allows for having engine power available as much as possible, from very-low-speed field operations up to their max speed for road transportation. The gearbox in this work was modelled using groups of simple gears connected in series or in parallel with the help of disengaging friction clutches to achieve the overall desired vehicle speed. Lastly, PTO loads representing the power required by the implements connected to the tractor were modelled through a resistive torque applied to the tractor PTO. The powers required by the implements were obtained in previous experimental field tests (Mocera et al. [22]) with the angular speed of the PTO fixed at 540 rpm. Table 2 shows a summary of these three working scenarios.

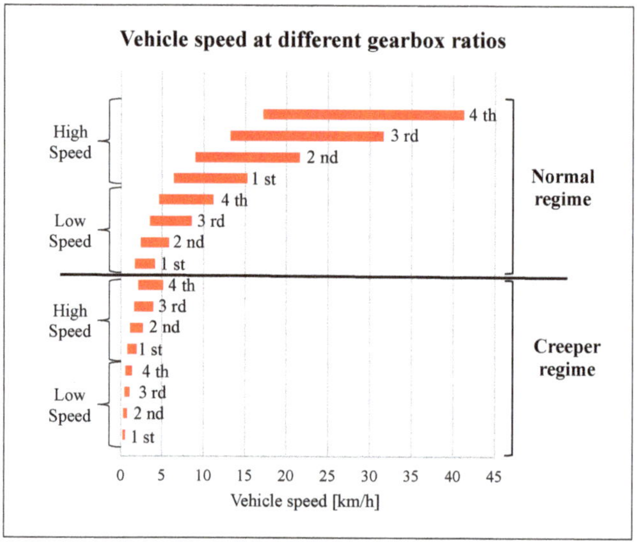

Figure 3. Typical agricultural gearbox configuration for maximal travel speed coverage.

Table 2. Relevant characteristics of proposed implement-related activities [22].

	Shredder	Atomizer	Rotart Harrow
Min power (kW)	12	32	36
Mean power (kW)	19	36	44
Max power (kW)	26	40	52
Vehicle speed (km/h)	5	5	5
ICE angular speed (rpm)	2244	2244	2244

3. Hybrid eCVT Powetrain

3.1. Architecture and Components

The proposed hybrid architecture consisted of a downsized thermal unit and two reversible electric machines. The main component of the transmission was a two-stage planetary gear that allowed for achieving a power-split hybrid configuration, and thereby electronic continuously variable transmission (eCVT). In detail, the ICE was mechanically connected through a clutch to the first electric machine (EM_1). The drive shaft was also linked to the sun gear of the second planetary gear stage and, with a clutch, to the power take off (PTO). The rotor of the second electric machine (EM_2) was mechanically connected to the sun gear of the first planetary gear stage, whose ring gear was fixed to the chassis. The carrier of this first stage was connected to the crown of the second planetary gear stage. Finally, the carrier of the second stage was connected to the drive shaft of a two speed gearbox which output were the rear and front axles differentials. Figure 4 shows the proposed powertrain layout. This configuration allowed for having no direct connection between ICE and wheels. However, with a proper control, it could allow for the joint cooperation with the electric machines to have full power at different vehicle speed levels. Through the disengagement of the clutch between ICE and EM_1, the powertrain was able to operate in full electric mode using the two electric machines together for the best use of the electric energy stored in the battery pack. Lastly, the position of EM_2 in this architecture enabled regenerative braking, although the impact was negligible if not for very special work scenarios.

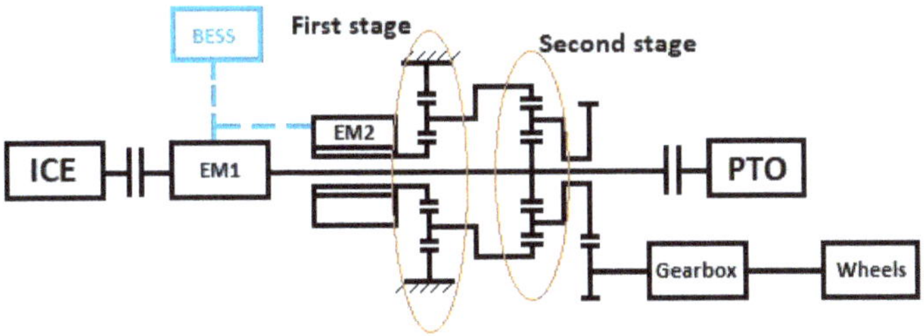

Figure 4. Schematic representation of HeCVT tractor architecture.

The use of a downsized ICE for this HeCVT tractor was motivated by the following considerations:

- Specialized tractors are usually equipped with high power diesel engines but, for a large part of their operative life, they work at medium–low load, which translates into suboptimal overall efficiency, and thus excessive fuel consumption. Instead, a downsized ICE working at higher percentage loads that are closer to the nominal design value can achieve better fuel efficiency.
- Tractors' emission limits are imposed by the NRMM regulation, which classifies engines according to their application and nominal rated power. The larger the engine, the higher the pollutant production per unit of work. Therefore, an oversized diesel engine has stricter restrictions about the pollutants that it is allowed to produce; thus, it would require large volumes for exhaust gas aftertreatment systems. This is not compatible with compact specialized tractors.

The power and torque curves of the downsized engine considered for the HeCVT tractor are shown in Figure 5. In this case, the proposed torque and power curves were also derived from data available online for commercial diesel engine units having nominal characteristics as close as possible to the desired ones.

The properties of the two electric machines are reported in Table 3. The numerical model of the EMs defines electrical losses as the sum of the following terms: fixed converter losses, ohmic losses (torque-dependent), and iron losses (speed-dependent) due to eddy currents. In some difficult situations, electric machines can count on a temporary boost function. It consisted of an increase in available torque due to the use of a higher level of current. However, this capability had the following constraints to preserve motor functionalities:

- at low speed below 1100 rpm, torque could be amplified up to twice the nominal value;
- above 1100 rpm, torque amplification should be limited by the rated power value for the machine.

Table 3. Main properties of proposed electric units.

EM_1 nominal power	35 kW
EM_1 maximal torque	145 Nm
EM_1 nominal efficiency	93%
EM_2 nominal power	30 kW
EM_2 maximal torque	130 Nm
EM_2 nominal efficiency	93%

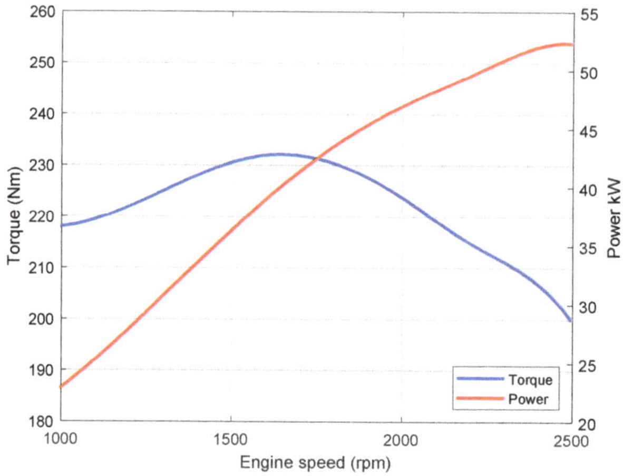

Figure 5. Torque and power characteristics for diesel engine of HeCVT tractor.

The main purpose of EM_2 was to control the vehicle speed, while the EM_1:

- helps the ICE in providing power to the PTO;
- is turned by the ICE in order to provide electric power to EM_2 for an electric power split;
- recirculates power when EM_2 acts as a brake for low vehicle speed, helping the ICE in parallel .

As mentioned before, the two-stage planetary gear allowed for considering a simplified two-speed gearbox with only two speed ranges to increase torque availability at low vehicle speed. The gear ratios were chosen so that the base speed of EM_2 reached approximately 15 km/h in a low-speed regime, and 40 km/h in a high-speed regime, with the ICE rotating at 2350 rpm. Table 4 shows the transmission ratios of the driveline.

Table 4. Transmission ratios of HeCVT powertrain.

Low speed ratio at gearbox	22.56
High speed ratio at gearbox	8.45
First-stage ring to sun teeth ratio	2
Second-stage ring to sun teeth ratio	2

Along with the electric machines and drives, an energy storage system consisting of a battery pack (BP) was introduced. To meet the typical compactness needs of specialized orchard tractors, the sizing of the BP capacity was chosen as a compromise between fuel consumption optimization and required volume. As in previous work [22], the proposed BP in this work had a nominal voltage of 600 V and a nominal capacity of 16 kWh. To meet the needs for a compact battery pack, lithium-ion technology was considered because of the higher energy density with respect to other chemistries available on the market. From a numerical point of view, the BP was modelled with an equivalent circuit model consisting of a voltage source and an internal resistance. The instantaneous electric power provided by the BESS was evaluated considering the power required by the EMs in accordance with a physical network (PN) modelling approach. The PN is a modular modelling approach where each element is considered to be a physical entity capable of exchanging energy with all the other subsystems to which it is connected, with the constraint that the solutions of the system must satisfy the power balance equations for each component. To evaluate the actual state of charge of the battery pack, the authors opted for a simple Coulomb counting strategy, neglecting more detailed models available in the literature. The following equations were used.

$$C_{used} = \int_{t_1}^{t_2} i_{batt} \quad dt \tag{6}$$

$$SOC = \frac{C_{nom} - C_{used}}{C_{nom}} \tag{7}$$

where:

- C_{used} is the used capacity;
- C_{nom} is the nominal capacity;
- SOC is the state of charge of the battery pack.

3.2. Hybrid eCVT Control

3.2.1. Kinematics

In order to properly control the powertrain, kinematic relations between different components and wheels must be explored. The notation used in the equations was based on the following principles:

- Subscripts S, C and R refer to sun, carrier, and ring gears, respectively.
- Subscripts 1 and 2 refer to first- and second-stage planetary gear, respectively.

In accordance to these premises, the following equations could be obtained:

$$i_{SR} = \frac{\omega_S - \omega_C}{\omega_R - \omega_C} \tag{8}$$

If $\omega_R = 0$, it is possible to assert:

$$i_{SC} = \frac{z_S + z_R}{z_S} \tag{9}$$

Therefore,

$$\omega_{C_1} = \frac{\omega_{S_1}}{i_{SC_1}} = \frac{\omega_{EM_2}}{1 + \frac{z_{R1}}{z_{S_1}}} = \omega_{R_2} \tag{10}$$

Regarding the second-stage planetary gear, it is possible to write:

$$\omega_{C_2} = \frac{\omega_{S_2} - i_{SR_2} \cdot \omega_{R_2}}{1 - i_{SR_2}} \tag{11}$$

Since $\omega_{S_2} = \omega_{ICE}$, it is valid that:

$$\omega_{EM_2} = \left[\omega_{C_2} * \left(1 + \frac{z_{R_2}}{z_{S_2}}\right) - \omega_{ICE}\right] * \frac{z_{S_2}}{z_{R_2}} * \left(1 + \frac{z_{R_1}}{z_{S_1}}\right) \tag{12}$$

where ω_{C_2} is kinematically linked to the wheels through the gearbox. Equation (12) enabled to control the EM_2 decoupling the rotational speed of the diesel engine from the actual vehicle speed. Figure 6 shows the values of EM_2 rotational speed and the second-stage carrier along the vehicle speed range of 0–40 km/h. At 15 km/h, the vehicle changed the speed range from low to high speed. Below a certain threshold speed for each regime, EM_2 is required to behave as a brake, while EM_1 can be used to recirculate power helping the ICE in this work range. Both electric machines could be used to charge the BP in the case of low SOC and low external load.

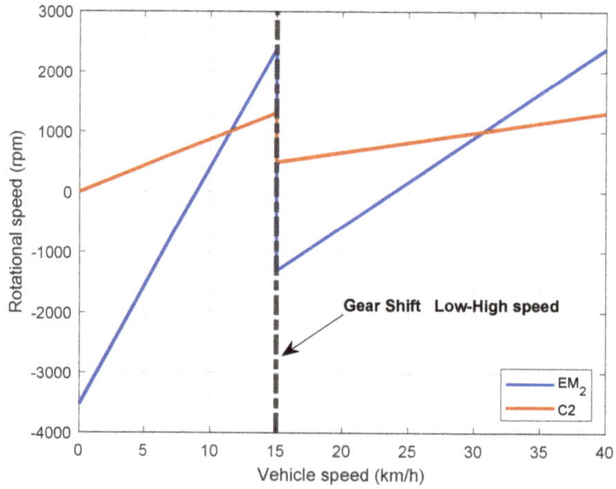

Figure 6. Kinematic relations between vehicle speed and EM_2 angular speed at fixed ICE rotational speed of 2350 rpm.

3.2.2. Energy Management

For this preliminary performance analysis, the proposed energy management strategy was based on two operating modes depending on the actual SOC of the battery pack: charge depleting (CD) mode and charge sustaining (CS) mode. When the battery SOC was above a certain threshold value, namely, 30% for this work, the powertrain operated in CD mode, optimizing fuel consumption; below, it operated in CS mode aiming to maintain the current state of charge value. Existing studies suggested that this threshold value should be around 55% to preserve battery life. However, other recent studies that considered a degradation model based on solid electrolyte interphase growth in a Li-ion battery cell showed that the optimal value for the charge sustaining mode is in the range of 36–38%. However, ccording to the same studies, 30% of SOC is a good value for CS mode [31]. In CD mode, the EM_1 torque control signal was obtained as the sum of two contributions deriving from the engine actual load and the battery current value. The first contribution was based on a function of the actual engine load (load observer [22]), and it depended on the sign and value of EM_2 rotational speed. When EM_2 speed was positive, EM_1 was mainly actuated for high values of the engine load (load observer), providing additional power to the sun gear drive shaft if necessary; when EM_2 speed was negative, acting as a brake or generator, EM_1 exploited the recirculation of power deriving from EM_2. As a consequence, EM_1 helped the ICE with a greater contribution. The second contribution that determined the EM_1 torque control signal was calculated as a function of the battery discharge current index, defined as the ratio between battery discharge current and maximal instantaneous acceptable discharge current, fixed at 5C (or five times the BP rated current). This contribution was equal to zero if the system needed to charge the BP. The main goal of the powertrain in CS mode was, if allowed by a low external load, to maintain the SOC level around 30%. When the requested power was too high, the powertrain preferred to accomplish the required task, for example, an occasional peak of power at the PTO and, as soon as possible, it recharged the batteries in order to bring the SOC level back to the threshold value. The EM_2 was instead controlled in accordance with the speed requested at the wheels. Lastly, the ICE, since it was not kinematically linked to the vehicle speed, was controlled setting a reference rotational speed depending on the specific work the tractor was tested against. As for regenerative braking, the control system operated using the EM_2 to reduce vehicle speed by applying a negative torque. While braking with EM_2, EM_1 was controlled to maintain its rotational speed and, as a consequence, that of the thermal unit at the reference value. During regenerative braking, the EM_2 first acted as a brake, since its angular speed was positive, and it was applying a negative torque. Then, it acted as a motor when its angular speed became negative. In this second case, the recirculation of power came from EM_1 acting as a brake to maintain the reference rotational speed constant. During regenerative braking, the torque of the two electric machines was controlled to not exceed the battery pack charge current limit, fixed at 1C. Figure 7 shows a schematic representation of the control strategy of the HeCVT architecture.

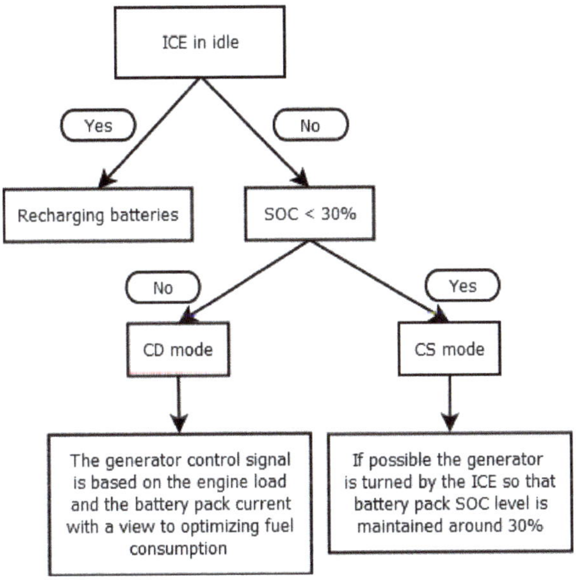

Figure 7. Energy management strategy for HeCVT tractor.

4. Numerical Investigation

For a comparison between the conventional powertrain and the proposed HeCVT, numerical simulations focused on performance and fuel consumption. As mentioned before, some constraints regarding maximal discharge and charge current at the battery pack were applied:

- Maximum battery pack current during continuous discharging equal to 3 C_{nom}.
- Maximum battery pack current during instantaneous discharging equal to 5 C_{nom} (max 10 s).
- Maximum battery pack current during continuous charging equal to 0.5 C_{nom}.
- Maximum battery pack current during instantaneous charging equal to 1 C_{nom}.

These constraints, derived from previous numerical and experimental activities available in the literature [18], were introduced to consider a hypothetical precautionary approach to preserve the state of health of the BP.

4.1. Acceleration Tests

The proposed acceleration tests aimed to evaluate the performance of the vehicle in the two most representative speed ranges with different payloads on the trailer. Therefore, the following tests were performed:

- 0–15 km/h with no trailer, with a trailer with 6000 kg and 10,000 kg of payloads.
- 0–40 km/h with no trailer, with a trailer with 6000 kg and 10,000 kg of payloads.

Table 5 shows the results of the acceleration tests, and highlights two different behaviors for the HeCVT in the test performed with a gear shift (GS) between the low and high speed ranges or in the case of a test started directly with the high speed range gear ratios (thus no GS). During the test with GS, the vehicle started in the low-speed regime and changed to a high-speed regime at 15 km/h. This comparison was proposed to show that, if the tractor did not have to pull a heavy trailer, it was possible to use directly the high speed regime without a consistent difference in performance. This would save clutches and other mechanical elements from unnecessary wear and would simplify the driving experience. Results for the conventional powertrain were obtained with gearshifts at 5, 10, and 20 km/h. The conventional powertrain model results were validated by the authors in a previous work using experimental data recorded during tractor field tests in different work scenarios [22,23]. In this work, the HeCVT architecture model was compared with the validated model of the conventional architecture with the same characteristics.

Table 5. HeCVT acceleration performance with different trailer payloads.

	HeCVT with GS	**HeCVT High Speed**	**Conventional**
0–15 km/h			
No trailer	1.2 s	2.1 s	2.2 s
6000 kg trailer	2.5 s	6.7 s	3.4 s
10,000 kg trailer	3.6 s	11.3 s	4.3 s
0–40 km/h			
No trailer	4.8 s	4.9 s	5.6 s
6000 kg trailer	14.5 s	17.8 s	14.7 s
10,000 kg trailer	23.1 s	30.3 s	24.4 s

The downsized diesel engine results showed comparable acceleration performance between the HeCVT and conventional tractors. The reason of this was in the possibility of the electric machines, in particular EM_2, to cooperate with the ICE through the two-stage planetary gear, providing traction power for the wheels. As an example, Figure 8 shows the simulation results of the 0–40 km/h acceleration test pulling a trailer of 10,000 kg (tare mass + payload). The first diagram shows the vehicle speed, the second the kinematic link between the ICE and the EM_2 rotational speed, the third the torque applied by each machine, and the last one their instantaneous power. In this test, it was possible to appreciate the transition from CD mode to CS mode at t = 33 s, when the power generated by EM_1 (negative sign) except for conversion losses was directly used by EM_2 to maintain the vehicle speed.

4.2. Slope Test

Another important test to evaluate the HeCVT tractor performance in terms of peak power capabilities is the slope test. In this case, different work scenarios were considered to explore different critical work conditions for the powertrain:

- Maximum approachable slope at 40 km/h with no trailer.
- Maximum approachable slope at 5 and 15 km/h with 10,000 and 5000 kg of payload, respectively.
- Maximum reachable speed with no trailer and 45% of slope.

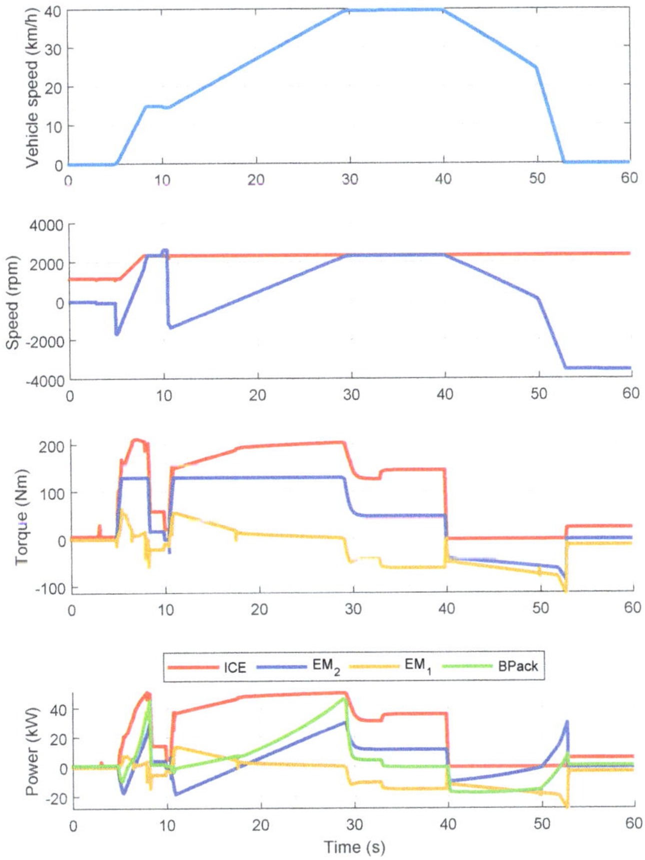

Figure 8. HeCVT 0–40 km/h acceleration test with 10,000 kg trailer and 0% of slope.

These tests aimed to verify the capability to provide full power at those different speed conditions under the constant load of the road slope. Moreover, each test was performed with a standing start at the prescribed slope to test the tractor capability to also start moving on the proposed slope condition (Italian traffic laws prescribe that the tractor should at least be able to stand start at 14% with the prescribed trailer load). Results are reported in Table 6. In almost all the work conditions, the HeCVT tractor slightly outperformed the conventional one except for the 5 km/h and 10,000 kg trailer test. In that case, the greater number of possible gear configurations given by a traditional gearbox configuration allowed for greater tractive effort at that specific speed, enabling the conventional ICE to work closer to its max power condition. Only in this special work scenario, the HeCVT system required the use of the boost function of the electric system in order to achieve the prescribed slope. The HeCVT tractor recorded good results at 40 and 15 km/h since its two speed regimes were dimensioned for those vehicle speeds.

Table 6. HeCVT performance in different slope tests. * Boost function activation required.

	HeCVT	Conventional
40 km/h, no trailer	25%	22%
5 km/h, 10,000 kg trailer	18% *	40%
15 km/h, 5000 kg trailer	20%	15%
Max speed 45% slope, no trailer	19.4 km/h	20.3 km/h

By way of example, Figure 9 shows the simulations results of the HeCVT architecture 0–15 km/h acceleration test with 5000 kg of payloads and 20% of slope.

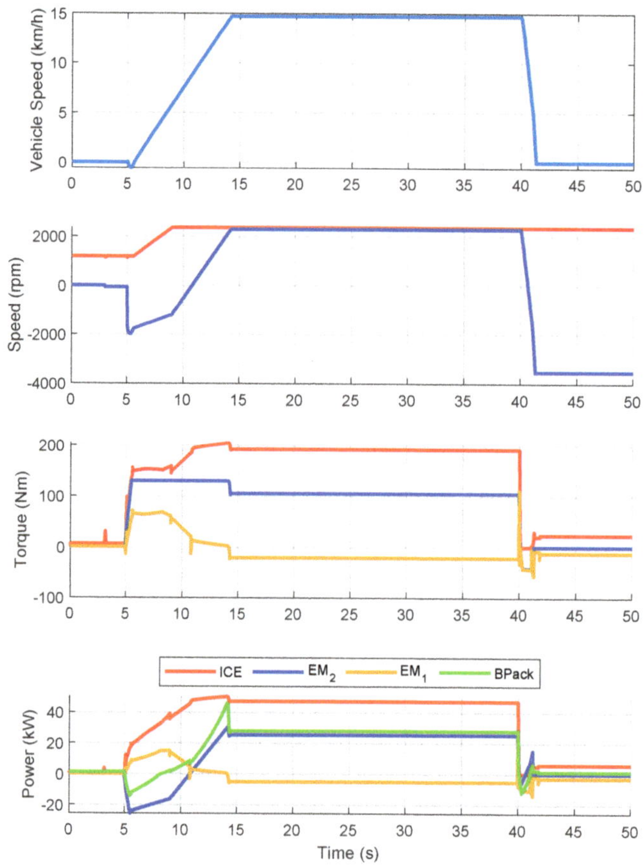

Figure 9. HeCVT 0–15 km/h acceleration test with 5000 kg trailer and 20% slope.

4.3. Work Cycles—Fuel Consumption

In this section, the fuel consumption of the conventional and HeCVT powertrain is compared. Numerical simulations were performed to evaluate the overall energy efficiency of the powertrains during work activities, in particular while pulling trailers and using attached implements powered by the PTO. In particular, the following aspects were investigated:

- Fuel and energy consumptions per hour in CD mode.
- Fuel consumption per hour in CS mode.
- Fuel consumption in proposed working cycles of 8 h.

The main parameters of PTO work activities are described in Table 2. Regarding CD mode, Table 7 shows the energy consumption, defined as the sum of the fuel equivalent energy content and the electric energy used from the battery pack. This value was compared in relative terms to the equivalent energy consumption of the conventional powertrain during the same tasks. For a fair comparison between the conventional and hybrid architecture, the equivalent energy approach was chosen to take into account the use of electric energy in the hybrid powertrain. Given the same mechanical work for the two case studies, the fuel consumption of the diesel engine alone would not be sufficient to demonstrate efficiency improvement because part of the requested energy would be satisfied by the electric reservoir. In this way, both sources were considered together in comparison with the only internal combustion engine of the conventional tractor. The HeCVT architecture recorded great energy saving during the trailer handling thanks to the possibility of optimizing the ICE working point, since the thermal unit was kinematically free from the wheels. Regarding CS mode, Table 7 shows the fuel consumption of the HeCVT architecture compared to the fuel consumption of the conventional powertrain of reference while performing the same tasks.

Table 7. Equivalent energy saving of the HeCVT architecture in CD and CS modes compared to the reference values of the conventional powertrain for the same tasks.

	CD Mode [%]	CS Mode [%]
Shredder	14	13
Atomizer	10	9
Rotary harrow	10	8
Handling at 40 km/h a 1400 kg trailer	30	25
Handling at 40 km/h a 6000 kg trailer	21	16
Handling at 40 km/h a 10,000 kg trailer	19	14

Lastly, two different types of work cycles were considered: PTO cycles and trailer handling. PTO cycles consisted of an 8 h work day: 90% of the time using the PTO powered implements and 10% of idling for accessory activities, where the tractor is left on to maintain the machine ready to be used. The trailer handling cycle consisted of an 8 h work day: for 45% of the time, handling was performed at 40 km/h pulling the full payload of the trailer (10,000 kg); for 45% of the time, handling was performed at 40 km/h with an empty trailer (1400 kg); for the remaining 10% of the time, the tractor was left idling. During tractor idling, the powertrain was controlled to recharge the batteries if necessary. Table 8 shows the results of these tests. In this case, daily fuel consumption was considered, comparing the HeCVT fuel consumption with that of the conventional tractor in the same test.

Table 8. Diesel fuel saving of the HeCVT architecture in the proposed 8 h work cycles compared to conventional diesel tractor consumption.

	HeCVT [%]
Shredder cycle	13
Atomizer cycle	10
Rotary harrow cycle	9
Handling trailer cycle	21

Results of these tests showed good improvement in energy and fuel consumption for the HeCVT architecture compared to that of the conventional powertrain. In particular, the hybrid powertrain achieved more than 20% of fuel saving in the handling trailer cycle, and an average of 10% of fuel saving in the PTO cycles. Figure 10 shows the results of a rotary harrow test, given as an example of the HeCVT powertrain behavior. Since the main purpose of this implement is to break up and smooth out the surface of the soil, the power requested at the PTO was simulated through the sum of a mean value and a ripple. In Figure 10, at t = 25 s the powertrain starts to operate in CS mode. In this mode, there is an increase in the EM2 power output irregularities. This is mainly due to the fact that, in CS mode, the powertrain has a stronger proportional controller, since it struggles to both preserve the BP SOC level if possible and to actuate the implements at 540 rpm.

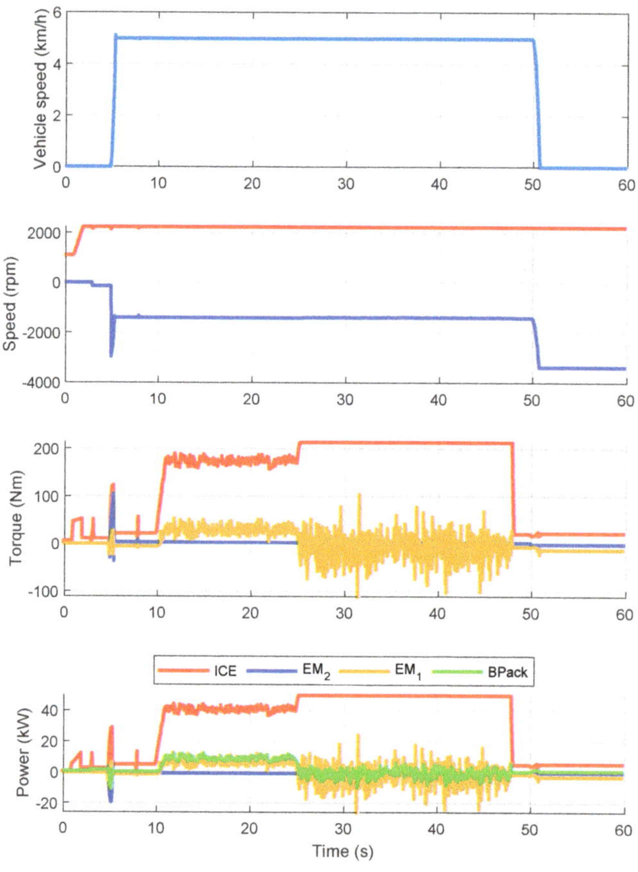

Figure 10. HeCVT rotary harrow test.

4.4. Full Electric Mode

The proposed HeCVT architecture allows for operating in full electric mode. This mode is available through the disengagement of the clutch between thermal unit and the drive shaft to which the generator is coupled. Full electric mode can be used in light work situation, since at high loads, the instantaneous discharging current of the battery pack would exceed the safety limit of 5C or endurance may be too low. To explore full electric capabilities, some of the work scenarios shown in previous sections were considered. Table 9 shows the powertrain performance in full electric mode, while Table 10 reports the endurance at different vehicle speed levels. Endurance means the time to discharge the battery SOC from 100% to 20%. Figure 11 shows the results of the 0–40 km/h acceleration in full electric mode with no trailer. When the vehicle reached speed of 15 and 40 km/h, the two electric machines had approximately the same rotational speed corresponding to their base speed. In addition, the power coming from the battery pack did not exceed either instantaneous or continuative previously imposed safety limits.

Table 9. HeCVT performance in full electric mode.

0–15 km/h no trailer	1.2 s
0–40 km/h no trailer	7 s
0–15 km/h 6000 kg trailer	3.7 s
0–40 km/h 6000 kg trailer	28.9 s
0-15 km/h 10,000 kg trailer	5.6 s
0–40 km/h 10,000 kg trailer	63.2 s
Max slope at 40 km/h no trailer	36%
Max slope at 15 km/h no trailer	10%

Table 10. HeCVT endurance in full electric mode: constant speed 40 km/h with different trailer payloads.

No trailer	66 min
6000 kg trailer	28 min
10,000 kg trailer	19 min

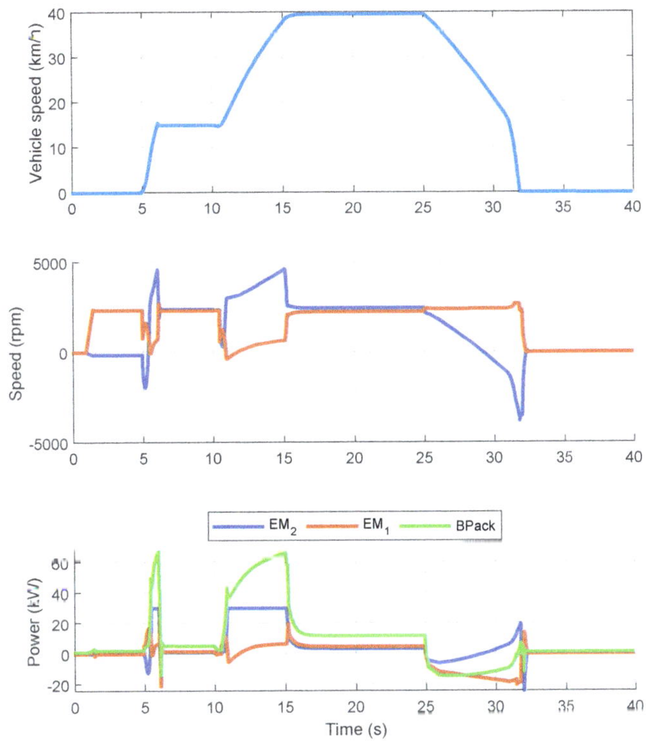

Figure 11. HeCVT 0–40 km/h acceleration test without trailer and 0% of slope in full electric mode.

5. Conclusions

In this work, a HeCVT tractor architecture was studied to investigate its performance in terms of peak power capabilities in field-derived work tasks and fuel saving in daily work cycles against a traditional orchard tractor taken as the case study. The HeCVT was equipped with a downsized diesel engine of 56 kW and two electric motors of 30 kW, whose capabilities were mechanically combined with a two-stage planetary gear. This layout allowed for both for ICE operating point optimization and to use the BP power reservoir as a booster thanks to the proper control of the two electric machines. The proposed HeCVT architecture showed promising results in almost all the test cases, achieving the best improvements in trailer handling where the conventional architecture is obliged to continuously change the engine rotational speed to follow the desired travel speed of the driver. Performance tests with several combinations of slopes and trailer payload were considered to see how the HeCVT powertrain was able transfer power to the wheels in different combinations of travel speed and resistive load. Results showed that the HeCVT was able to perform in the acceleration and overcoming of slopes as much as the conventional powertrain in almost all the performed tests. Therefore, the proposed hybrid architecture is a valid alternative to the traditional one since it achieved at least the same performance in terms of peak power capabilities, but is also able to accomplish typical daily work activities with a significant reduction in fuel consumption. Moreover, the compact longitudinal design of the transmission and the size of the proposed battery pack are suitable for the required compactness in orchard tractors. Future works will explore further optimization techniques for ICE operating points and in-depth design optimization of powertrain elements without compromising the good results achieved with this preliminary solution.

Author Contributions: Conceptualization, F.M. and V.M.; methodology, F.M. and V.M.; software, F.M. and V.M.; validation, F.M. and V.M.; formal analysis, F.M. and V.M.; investigation, F.M. and V.M.; resources, F.M.; data curation, F.M. and V.M.; writing—original draft preparation, F.M. and V.M.; writing—review and editing, F.M. and V.M.; visualization, F.M. and V.M.; supervision, F.M.; project administration, F.M.; funding acquisition, F.M. All authors have read and agreed to the published version of the manuscript.

Funding: This work was supported in part by the Italian MISE project Brevetti+—Ecothea (C65F20000290008).

Institutional Review Board Statement: Not applicable.

Informed Consent Statement: Not applicable.

Data Availability Statement: Not applicable.

Acknowledgments: The authors would like to thank Ecothea Srl for the possibility to explore the full capabilities of the patented eCVT architecture thanks to funds from Italian MISE project Brevetti+.

Conflicts of Interest: The authors declare no conflict of interest.

References

1. Gorjian, S.; Ebadi, H.; Trommsdorff, M.; Sharon, H.; Demant, M.; Schindele, S. The advent of modern solar-powered electric agricultural machinery: A solution for sustainable farm operations. *J. Clean. Prod.* **2021**, *292*, 126030. [CrossRef]
2. Platis, D.P.; Anagnostopoulos, C.D.; Tsaboula, A.D.; Menexes, G.C.; Kalburtji, K.L.; Mamolos, A.P. Energy Analysis, and Carbon and Water Footprint for Environmentally Friendly Farming Practices in Agroecosystems and Agroforestry. *Sustainability* **2019**, *11*, 1664. [CrossRef]
3. Lovarelli, D.; Bacenetti, J. Exhaust gases emissions from agricultural tractors: State of the art and future perspectives for machinery operators. *Biosyst. Eng.* **2019**, *186*, 204–213. [CrossRef]
4. Bacenetti, J.; Lovarelli, D.; Facchinetti, D.; Pessina, D. An environmental comparison of techniques to reduce pollutants emissions related to agricultural tractors. *Biosyst. Eng.* **2018**, *171*, 30–40. [CrossRef]
5. European Parliament—Council of the European Union. Regulation (EU). 2016/1628 of the European Parliament and of the Council of 14 September 2016 on requirements relating to gaseous and particulate pollutant emission limits and type-approval for internal combustion engines for non-road mobile machinery, amending Regulations (EU) No 1024/2012 and (EU) No 167/2013, and amending and repealing Directive 97/68/EC. *Off. J. Eur. Union* **2016**, *252*, 53–117.
6. Scolaro, E.; Beligoj, M.; Estevez, M.P.; Alberti, L.; Renzi, M.; Mattetti, M. Electrification of Agricultural Machinery: A Review. *IEEE Access* **2021**, *9*, 164520–164541. [CrossRef]
7. Beltrami, D.; Iora, P.; Tribioli, L.; Uberti, S. Electrification of compact off-highway vehicles—Overview of the current state of the art and trends. *Energies* **2021**, *14*, 5565. [CrossRef]

8. Moreda, G.; Muñoz-García, M.; Barreiro, P. High voltage electrification of tractor and agricultural machinery—A review. *Energy Convers. Manag.* **2016**, *115*, 117–131. [CrossRef]
9. Somà, A. *Trends and Hybridization Factor for Heavy-Duty Working Vehicles*; IntechOpen: Rijeka, Croatia, 2017; Chapter 1 . [CrossRef]
10. Somà, A.; Bruzzese, F.; Mocera, F.; Viglietti, E. Hybridization Factor and Performance of Hybrid Electric Telehandler Vehicle. *IEEE Trans. Ind. Appl.* **2016**, *52*, 5130–5138. [CrossRef]
11. DLG-PowerMix: Tractor Output, Efficiency and Fuel Consumption. Available online: www.dlg.org/en/agriculture/tests/dlg-powermix (accessed on 26 January 2022).
12. Somà, A.; Mocera, F.; Bruzzese, F.; Viglietti, E. Simulation of dynamic performances of electric-hybrid heavy working vehicles. In Proceedings of the 2016 Eleventh International Conference on Ecological Vehicles and Renewable Energies (EVER), Monte Carlo, Monaco, 6–8 April 2016; pp. 1–8. [CrossRef]
13. Mocera, F.; Somà, A. Working Cycle requirements for an electrified architecture of a vertical feed mixer vehicle. *Procedia Struct. Integr.* **2018**, *12*, 213–223. [CrossRef]
14. Lombardi, G.V.; Berni, R. Renewable energy in agriculture: Farmers willingness-to-pay for a photovoltaic electric farm tractor. *J. Clean. Prod.* **2021**, *313*, 127520. [CrossRef]
15. Renius, K.T. *Fundamentals of Tractor Design*; Springer International Publishing: Cham, Switzerland, 2020. [CrossRef]
16. Liu, M.; Xu, L.; Zhou, Z. Design of a Load Torque Based Control Strategy for Improving Electric Tractor Motor Energy Conversion Efficiency. *Math. Probl. Eng.* **2016**, *2016*, 1–14. [CrossRef]
17. Li, T.; Xie, B.; Li, Z.; Li, J. Design and Optimization of a Dual-Input Coupling Powertrain System: A Case Study for Electric Tractors. *Appl. Sci.* **2020**, *10*, 1608. [CrossRef]
18. Mocera, F.; Vergori, E.; Soma, A. Battery Performance Analysis for Working Vehicle Applications. *IEEE Trans. Ind. Appl.* **2020**, *56*, 644–653. [CrossRef]
19. Lagnelöv, O.; Dhillon, S.; Larsson, G.; Nilsson, D.; Larsolle, A.; Hansson, P.A. Cost analysis of autonomous battery electric field tractors in agriculture. *Biosyst. Eng.* **2021**, *204*, 358–376. [CrossRef]
20. Mocera, F.; Somà, A. A Review of Hybrid Electric Architectures in Construction, Handling and Agriculture Machines. In *New Perspectives on Electric Vehicles*; IntechOpen: London, UK, 2021. [CrossRef]
21. Dalboni, M.; Santarelli, P.; Patroncini, P.; Soldati, A.; Concari, C.; Lusignani, D. Electrification of a Compact Agricultural Tractor: A Successful Case Study. In Proceedings of the 2019 IEEE Transportation Electrification Conference and Expo (ITEC), Novi, MI, USA, 19–21 June 2019; IEEE: Piscataway, NJ, USA, 2019; pp. 1–6. [CrossRef]
22. Mocera, F.; Somà, A. Analysis of a Parallel Hybrid Electric Tractor for Agricultural Applications. *Energies* **2020**, *13*, 3055. [CrossRef]
23. Mocera, F. A Model-Based Design Approach for a Parallel Hybrid Electric Tractor Energy Management Strategy Using Hardware in the Loop Technique. *Vehicles* **2020**, *3*, 1–19. [CrossRef]
24. Baek, S.Y.; Kim, Y.S.; Kim, W.S.; Baek, S.M.; Kim, Y.J. Development and Verification of a Simulation Model for 120 kW Class Electric AWD (All-Wheel-Drive) Tractor during Driving Operation. *Energies* **2020**, *13*, 2422. [CrossRef]
25. Jia, C.; Qiao, W.; Qu, L. Numerical Methods for Optimal Control of Hybrid Electric Agricultural Tractors. In Proceedings of the 2019 IEEE Transportation Electrification Conference and Expo (ITEC), Novi, MI, USA, 19–21 June 2019; IEEE: Piscataway, NJ, USA, 2019; pp. 1–6. [CrossRef]
26. Jia, C.; Qiao, W.; Qu, L. Modeling and Control of Hybrid Electric Vehicles: A Case Study for Agricultural Tractors. In Proceedings of the 2018 IEEE Vehicle Power and Propulsion Conference (VPPC), Chicago, IL, USA, 27–30 August 2018; IEEE: Piscataway, NJ, USA, 2018; pp. 1–6. [CrossRef]
27. Rossi, C.; Pontara, D.; Falcomer, C.; Bertoldi, M.; Mandrioli, R. A Hybrid–Electric Driveline for Agricultural Tractors Based on an e-CVT Power-Split Transmission. *Energies* **2021**, *14*, 6912. [CrossRef]
28. Miller, J.M. Hybrid Electric Vehicle Propulsion System Architectures of the e-CVT Type. *IEEE Trans. Power Electron.* **2006**, *21*, 756–767. [CrossRef]
29. Tebaldi, D.; Zanasi, R. Modeling control and simulation of a parallel hybrid agricultural tractor. In Proceedings of the 2021 29th Mediterranean Conference on Control and Automation, MED 2021, Bari, Italy, 22–25 June 2021; pp. 317–323. [CrossRef]
30. Somà, A. Hybrid Transmission Unit for a Tractor and Tractor Comprising the Same. WO2021234758A1, 18 May 2020.
31. Zhang, S.; Zhang, J. Optimal State-of-Charge Value for Charge-Sustaining Mode of Plug-In Hybrid Electric Vehicles. *IEEE Access* **2020**, *8*, 187959–187964. [CrossRef]
32. Grisso, R.B. *Predicting Tractor Diesel Fuel Consumption*; Publication 442-073; Virginia Cooperative Extension: Blacksburg, VA, USA, 2011.

applied sciences

MDPI

Article

Power Losses Minimization for Optimal Operating Maps in Power-Split HEVs: A Case Study on the Chevrolet Volt

Antonella Castellano and Marco Cammalleri *

Department of Engineering, University of Palermo, 90128 Palermo, Italy; antonella.castellano@unipa.it
* Correspondence: marco.cammalleri@unipa.it

Abstract: The power-split architecture is the most promising hybrid electric powertrain. However, a real advantage in energy saving while maintaining high performance can be achieved only by the implementation of a proper energy management strategy. This requires an optimized functional design before and a comprehensive analysis of the powertrain losses after, which could be rather challenging owing to the constructive complexity of the power-split transmission, especially for multi-mode architecture with multiple planetary gearing. This difficulty was overcome by a dimensionless model, already available in the literature, that enables the analysis of any power-split transmission, even in full electric operation. This paper relies on this approach to find the operating points of the internal combustion engine and both electric machines which minimize the total power losses. This optimization is carried out for given vehicle speed and demanded torque, by supposing different scenarios in respect of the battery capability of providing or gathering power. The efficiency of the thermal engine and the electric machines is considered, as well as the transmission mechanical power losses. The aim is to provide a global efficiency map that can be exploited to extract data for the implementation of the most suitable real-time control strategy. As a case study, the procedure is applied to the multi-mode power-split system of the Chevrolet Volt.

Keywords: hybrid electric vehicles; power-split powertrain; global efficiency; multi-mode transmission; Voltec analysis

Citation: Castellano, A.; Cammalleri, M. Power Losses Minimization for Optimal Operating Maps in Power-Split HEVs: A Case Study on the Chevrolet Volt. *Appl. Sci.* **2021**, *11*, 7779. https://doi.org/10.3390/app11177779

Academic Editor: Maria Grazia De Giorgi

Received: 29 July 2021
Accepted: 19 August 2021
Published: 24 August 2021

1. Introduction

In the last decades, stricter environmental policies undertaken against increasing global warming have encouraged the spreading uptake of hybrid electric vehicles. Besides the earliest and most popular series and parallel hybrid architectures, more and more automotive companies—first and foremost Toyota and General Motors—are developing the power-split hybrid electric powertrain [1–5].

The power-split layout combines the benefits of both series and parallel hybrid, resulting in a highly flexible system where the internal combustion engine (ICE) is kinematically decoupled by the wheels due to the operation of the electric unit. Two electric machines act as an active continuously variable unit (CVU), which can provide additional power for vehicle propulsion or gather the ICE power in surplus for battery recharging. In addition, regenerative braking and full electric vehicle (FEV) operation are achievable. The power flows within the powertrain are handled by a power-split unit (PSU) made up of planetary gear sets (PGs) and ordinary gear sets (OGs). Multi-PG PSU enables the minimization of the electric machines' power size by deploying a multi-mode power-split continuously variable transmission (PS-CVT), where some clutches operations lead to several constructive arrangements to select under the desired driving conditions [6–10]. Nevertheless, the more complex the transmission constructive layout is, the trickiest the identification of the occurring power flows is [11–13], as well as their management [14–17].

When implementing energy management strategies aimed at minimizing the powertrain power losses, the friction losses occurring in the transmission should be also considered. Nonetheless, because of the above-mentioned difficulties in multi-mode PS-CVTs analysis, their calculation is far from trivial [18–21]. Thus, mechanical power losses are considered in very simple one-PG transmission [22,23] or evaluated by using simplified approaches that avoid the analysis of the transmission power flows [24,25] or even more often neglected [26–32].

The focus of this paper is to provide the global efficiency map of the multi-mode Chevrolet Volt hybrid electric powertrain by assessing all powertrain losses, not only those occurring in the propulsors but also the transmission mechanical power losses. This can be achieved by applying a unified parametric model [33–36] without convoluted or case-specific formulations. All the relationships of the model are ruled by few basic functional parameters which can be swiftly derived from any PS-CVT constructive arrangement, even multi-PG or multi-mode [34,36]. These enable the calculation of speed, torque, and power ratios in real conditions, by using a fast black-box method [35,36] to assess the PSU mechanical losses and then the actual power that the electric machines should provide to the PSU for given ICE power and required output. The whole procedure applies also to the FEV operation [36].

This unified parametric model was first applied to the transmission of the Chevrolet Volt in [34], where an alternative design was proposed but any power losses were neglected. Then, the mechanical losses occurring in the PSU were considered in the Chevrolet Volt dimensionless analysis carried out in [35]. This article aims to consider also the power losses occurring in the propulsors, by moving from the dimensionless variables to those expressed in physical units of measurement. In addition, the analysis of the Chevrolet Volt in FEV operation is presented for the first time.

The utilization of dimensionless independent and dependent variables expressed as speeds, torques, or powers ratios results in utmost generality, which makes the model suitable for analyzing any PS-CVT. Nonetheless, the loss factors of the propulsors are strictly related to their operating point, therefore the brake specific fuel consumption (BSFC) map of the engine and the efficiency maps of the electric machines have to be introduced. By considering different combinations of input and output speeds and torques, it is possible to investigate the powertrain response in terms of the actual functioning point of the propulsors and related power losses, net power flow provided or gathered by the battery, and powertrain global efficiency. This approach outputs the results in some optimal operating maps collecting all data of interest, which allows the determination of the powertrain optimal functioning points based on the desired goal of the energy management strategy to implement. By way of example, this contribution proposes a simplified control strategy in steady-state driving by taking into consideration four possible state of charges (SOCs) of the battery. However, this approach is the ideal basis for the development of new control and energy management strategies, as it applies to any hybrid electric powertrain without requiring an in-deep knowledge of the behavior of each component of the powertrain, thanks to its generality. Therefore, it offers a neutral environment for engineers with different expertise.

The paper is organized as follows. Section 2 includes the dimensionless PSU mechanical power losses and the dimensionless speeds and real powers of the electric machines in the Chevrolet Volt powertrain in power-split operation, which were previously calculated in [35]. Then, the same variables are evaluated in FEV operation by using the procedure presented in [36]. In Section 3, the propulsors efficiency maps are introduced and the method for calculating and selecting the optimal operating points is described. Section 4 provides and discusses the results of the application, while Section 5 concludes the article by framing the new contribution in a broader context.

2. Dimensionless Parametric Approach for Voltec Analysis

The transmission system of the Chevrolet Volt, the so-called Voltec, is a PS-CVT with two PGs and three clutches which enable the multi-mode functioning. A third PG combined with a chain drive acts as a fixed-ratio OG in the final drive. Figure 1 shows the Voltec functional layout derived from [37] and previously described in [34,35]. The PGs and the final drive make up the PSU, which can be considered as a four-port device linked with the ICE (by the shaft *in*), the wheels (by the shaft *out*), and the electric motor-generator (MG) I (by the shaft *i*) and MG O (by the shaft *o*). The positive sign of the power flows is indicated by the arrows. The clutches C0, C1, and C2 are exploited to shift between different modes, as reported in Table 1. By engaging the only C2 clutch, the PG2 ring gear is braked to the frame, thus only PG1 acts as an epicyclic gear unit with non-proportional speeds of its branches. This realizes an input-split mode, mainly exploited for lower vehicle speeds. At higher speeds, a compound-split mode is achieved, by engaging only the clutch C1 which connects the PG2 ring gear to the MG I and the PG1 sun gear. It should be noted that engaging C1 and C2 simultaneously realizes a fixed-ratio parallel mode, where only MG O can be active and the ICE speed is univocally coupled to the vehicle speed. Moreover, by additionally engaging the one-way clutch C0, which locks to the frame the ICE and the PG1 ring gear, two FEV modes can be performed. However, as shown in Table 1, General Motors considers only the FEV operation derived from the input-split arrangement.

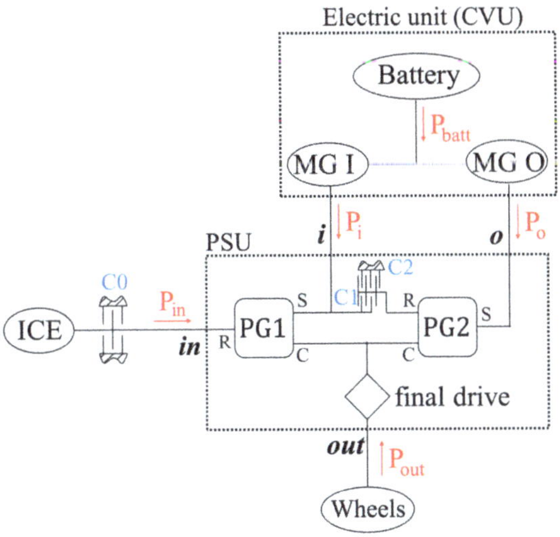

Figure 1. Functional layout of Voltec. The PGs ring gear, sun gear, and carrier are indicated by R, S, and C, respectively. Red arrows indicate the positive direction of power flows. Clutches are indicated by blue labels.

Table 1. Clutches operations for Voltec operating modes provided by General Motors.

Mode	C0	C1	C2
Input-Split			X
Compound-Split		X	
FEV	X		X

The schematization of any PSU as a four-port device is always valid, regardless of the actual PSU constructive layout. This enables the exploitation of some general relationships between speeds, torques, and powers of the main PSU external ports which can be applied to any PS-CVT [33,35,36]. On the other hand, the PGs and OGs constructive parameters and their arrangement within the PSU lead to the definition of the basic functional parameters which rule the equations of the unified parametric model considered in this article. The constructive parameter here used for the definition of the planetary gear sets is the Willis' ratio Ψ, defined as the ratio between the rotational speed of the ring gear and the one of the sun gear while the carrier is still. The Willis' ratios of the two PGs are $\Psi_1 = -0.535$ and

$\Psi_2 = -0.481$, the fixed-ratio of the final drive is $k_{fd} = 0.379$. The functional parameters of the Voltec input-split and compound-split modes were identified in [34]. These are the mechanical points $\tau_{\#i}$ and $\tau_{\#o}$ and the corresponding speed ratios $\tau_{o\#i}$ and $\tau_{i\#o}$, listed in Table 2. The former are defined as the overall speed ratio $\tau = \omega_{out}/\omega_{in}$ achieved when the i or o shaft is motionless, respectively. In general terms, the corresponding speed ratio $\tau_{j\#k}$ is the j-th speed ratio $\tau_j = \omega_j/\omega_{in}$ achieved when the shaft k is motionless. The mechanical points often coincide with the overall speed ratio at which a mode shift occurs, since one of the two electric machines can be turned off. Therefore, a parallel hybrid functioning is achieved at the mechanical points.

Table 2. Basic functional parameters of Voltec derived in [34].

Mode	$\tau_{\#i}$	$\tau_{\#o}$	$\tau_{o\#i}$	$\tau_{i\#o}$
Input-Split	0.247	0	2.00	−1.87
Compound-Split	0.247	0.510	2.00	2.00

Once known the parameters of Table 2, the dimensionless approach addressed in [35,36] can be applied to the Voltec to analyze both the power-split and FEV operation in terms of speed, torque, and power ratios, by including the evaluation of the PSU mechanical power losses.

2.1. Dimensionless Speeds, Powers and Mechanical Losses in Voltec Power-Split Operation

The Voltec PS-CVT in power-split operation was previously analyzed in [34,35]. To avoid repetition, this section includes only the outcomes of these previous applications, not the procedure implemented to obtain them. The results are shown in Figure 2. Starting from the functional parameters of Table 2, the speed ratio $\tau_i = \omega_i/\omega_{in}$ between MG I and the ICE was computed as a function of the overall speed ratio τ, as well as the speed ratio $\tau_o = \omega_o/\omega_{in}$ between MG O and the ICE. These are shown in Figure 2a for both input- and compound-split mode. The shift between one mode to the other occurs at the mechanical point $\tau = \tau_{\#i} = 0.247$. For $\tau = \tau_* = 0.379$ both electric machines rotate at the same speed, therefore both PGs work at their synchronous condition. At the PGs synchronism, the PSU mechanical power losses (Figure 2b) show a minimum, because the absence of relative motion between PGs branches eliminates the PGs friction losses. The mechanical power losses of Figure 2b were calculated as a fraction of the input power as a function of the overall speed ratio τ and the opposite of the overall power ratio $\eta = -P_{out}/P_{in}$. Note that η is not a global efficiency, but a variable exploited to model the possibility of the battery to provide or absorb power. Therefore, it can also be far higher than one, if the demanded output power is mainly provided by the battery rather than the ICE. The PSU losses enabled the calculation of the real power that the electric machines should provide to or collect from the PSU as a fraction of the input power (Figure 2c,d).

2.2. Dimensionless Speeds, Powers and Mechanical Losses in Voltec Full-Electric Operation

The relationships exploited in [34,35] to analyze the power-split operation were rearranged in [36] to model also the FEV functioning mode. These are exploited in this section to apply the unified parametric model to the Voltec in FEV operation for the first time. Nonetheless, to find the best balance between brevity and self-consistency of the paper, this section presents the only implementation of formulas that were introduced and explained in the previous works [33–36], to which we refer the reader for more information. However, Appendix A outlines the major features of the model to facilitate the understanding of the content of this section.

As the engine is inactive in FEV driving (clutch C0 is engaged), speeds and power can be more conveniently normalized to the output ones. Since the shaft *in* is motionless, the speed ratio between every electric machine and the shaft *out* is univocally defined as:

$$\begin{aligned} \left.\frac{\omega_i}{\omega_{out}}\right|_{\omega_{in}=0} &= \frac{\tau_{i_{\#o}}}{\tau_{\#o}-\tau_{\#i}} = 7.58 \\ \left.\frac{\omega_o}{\omega_{out}}\right|_{\omega_{in}=0} &= \frac{\tau_{o_{\#i}}}{\tau_{\#i}-\tau_{\#o}} = 8.12 \end{aligned} \tag{1}$$

The functional parameters used in Equation (1) are those related to the input-split mode, since it is the only mode exploited by General Motors to perform the FEV operation.

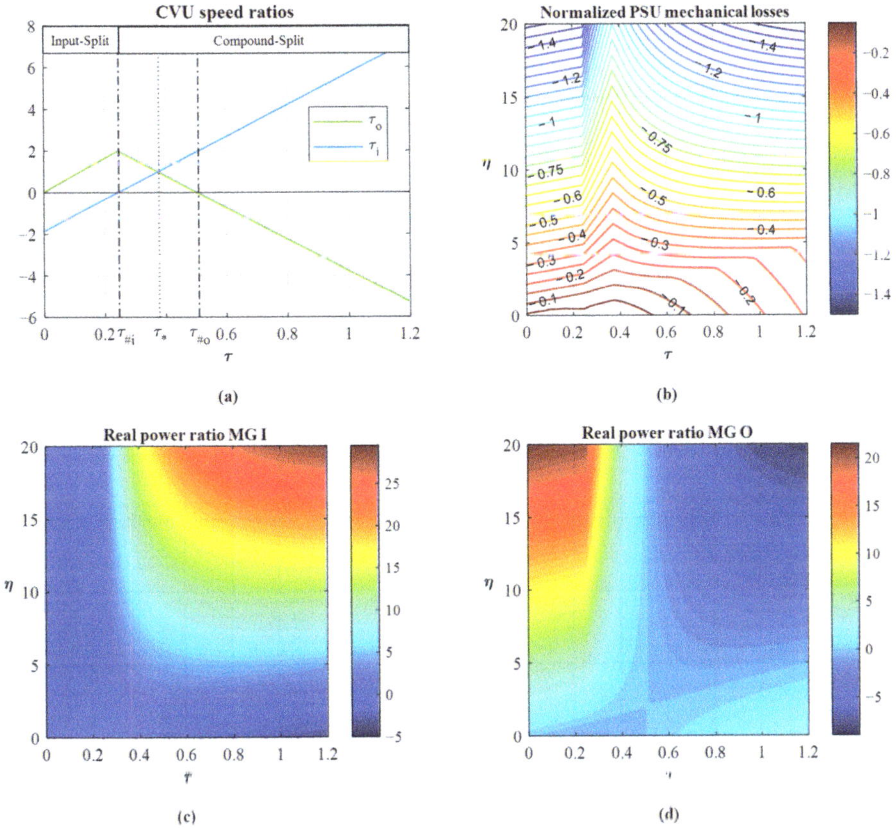

Figure 2. Results of Voltec analysis in power-split operation carried out in [34,35]. (**a**) CVU speed ratios; (**b**) mechanical power losses in the PSU as a fraction of the input power; (**c**) MG I real power as a fraction of the input power; (**d**) MG O real power as a fraction of the input power.

The dimensionless PSU power losses can be computed by the fast black-box method proposed in [35] adapted to the FEV analysis, as described in [36] and summarized in Appendix A. The considered efficiencies of the final drive ($\eta_{fd} = 0.953$) and of the PGs in fixed-carrier functioning ($\eta_0 = 0.96$) are the same as those assumed in [35]. The total mechanical power losses are the sum of those occurring in the final drive, calculated as:

$$\left.\overline{p}'_L\right|_{OG} = -\frac{\left.\overline{P}_{Loss}\right|_{OG}}{P_{out}} \approx -\left|\left(1-\eta_{fd}\right) p'_{out}\right| \tag{2}$$

and those occurring in the PGs:

$$\overline{p}'_L\big|_{PG1} = -\frac{\overline{P}_{Loss}\big|_{PG1}}{P_{out}} \approx -\left|\left(1-\eta^R\right)\left(\frac{\phi^{in}_{out/i} - \psi^R_{C/S}}{1-\psi^R_{C/S}}\right)p'_{out,1}\right| \tag{3}$$

$$\overline{p}'_L\big|_{PG2} = -\frac{\overline{P}_{Loss}\big|_{PG2}}{P_{out}} \approx -\left|\left(1-\eta^C\right)\left(\frac{\phi^{out}_{o/i} - \psi^C_{S/R}}{1-\psi^C_{S/R}}\right)p'_o\right| \tag{4}$$

The parameters used in Equations (3) and (4) are indicated in Table 3, while in Equation (3) $p'_{out,1}$ is the portion of p'_{out} flowing in PG1, which can be computed as the difference between the power flowing into the final drive (p'_{out}/η_{fd}) and its portion flowing in PG2 ($p'_{out,2}$, see Equation (A4) in Appendix A for its calculation), as follows:

$$p'_{out,1} = \frac{p'_{out}}{\eta_{fd}} - p'_{out,2} = \frac{p'_{out}}{\eta_{fd}} + p'_o\phi^i_{o/out} \tag{5}$$

Table 3. PGs reference notation, fixed-Z speed ratios, and fixed-Z efficiency [35,36].

	$x-X$	$y-Y$	$z-Z$	$\psi^Z_{X/Y}$	η^Z
PG1	*out* – Carrier	*i* – Sun	*in* – Ring	$\psi^R_{C/S} = \frac{\Psi_1}{\Psi_1-1}$	$\eta^R = \frac{1-\Psi_1}{\eta_0-\Psi_1}$
PG2	*o* – Sun	*i* – Ring	*out* – Carrier	$\psi^C_{S/R} = \frac{1}{\Psi_2}$	$\eta^C = \eta_0$

Note that $\overline{p}'_{out} = p'_{out} = -P_{out}/P_{out} = -1$ by definition. By considering $\overline{p}'_o = p'_o = -P_o/P_{out}$ as the independent variable, the total power losses $\overline{p}'_L$ can be swiftly computed by summing Equations (2)–(4). The dimensionless power flowing in the other electric machines can be calculated by the PSU real power balance:

$$\overline{p}'_o + \overline{p}'_i + \overline{p}'_L = 1 \tag{6}$$

The results of the dimensionless analysis to FEV operation are shown in Figure 3.

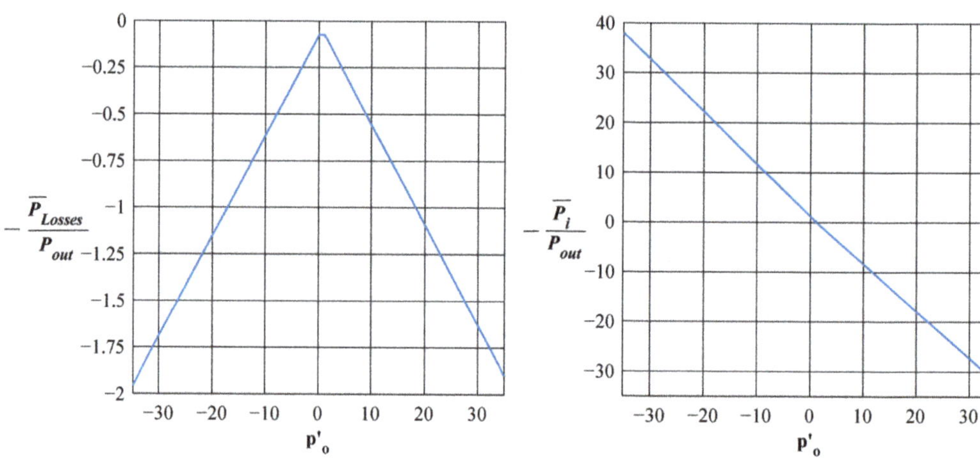

(a) (b)

Figure 3. Results of Voltec analysis in FEV operation. (**a**) Mechanical power losses in the PSU as a fraction of $-P_{out}$; (**b**) MG I real power as a fraction of $-P_{out}$.

3. Identification of the Optimal Operating Maps

To carry out the analysis addressed in Section 2, it is sufficient to know only the constructive and functional layout of the PS-CVT. This allows the calculation of the PSU speed ratios as well as the dimensionless mechanical power that electric machines should provide or absorb, by considering the PSU friction losses. More specifically, in power-split operation, the dependent variables, which are normalized to ICE speed or power, can be assessed by freely assigning the speed ratio and the power ratio between the output and the input port (Figure 2). On the other hand, in FEV operation, the speed ratios are univocally defined, thus the dependent variables, which are normalized to the output power, can be determined by freely supposing the power ratio between one PSU port connected to an electric machine and the output port (Figure 3). In other words, for a given vehicle speed (directly related to ω_{out}) and for a given demanded torque, it is necessary to assume a specific functioning point of the ICE (in PS operation) or of one electric machine (in FEV operation) to univocally determine speeds and powers (and thus torques) on each PSU port.

Obviously, the output torque and the wheels' speed can be considered independently one from the other, but if the output torque is higher or lower than the driving resistance, the powertrain work in dynamic operations, and the inertia of the vehicle and the propulsors should be considered in the analysis. In this article, for computational simplicity, the steady-state operation is analyzed, whereby the output torque is a function of the vehicle speed (V_{veh}), thus the power delivered by the powertrain is:

$$P_{out} = -\left(mg \sin\alpha + f_r mg \cos\alpha + 0.5 \cdot C_d A_f \rho_a V_{veh}^2\right) V_{veh} \tag{7}$$

where m, f_r, C_d, A_f are the Chevrolet Volt parameters reported in Table 4, α is the road slope in radians, $g = 9.81\ m/s^2$ is the gravitational acceleration, and $\rho_a = 1.225\ kg/m^3$ is the air density. It is worth noting that m is the sum of the unladen vehicle mass m_0 and the mass of passengers, fuel, or any other additional load. Therefore, m and α depend on the driving conditions. Nonetheless, it is sufficient to modify these values upstream of the procedure described in this section so as to obtain data referred to any steady-state driving condition.

Table 4. Chevrolet Volt parameters.

Unladen Mass m_0 [kg]	Rolling Resistance f_r	Drag Coefficient C_d	Frontal Area A_f [m^2]	Wheel Radius R_w [m]
1603	0.0113	0.28	2.20	0.323

Once fixed a vehicle speed, the functioning point of the ICE in power-split operation or of one electric machine in FEV operation has to be selected so as to determine $\tau = V_{veh}/(R_w \cdot \omega_{in})$ and $\eta = -P_{out}/P_{in}$, or $p_o' = -P_o/P_{out}$, respectively. To investigate all the viable powertrain functioning points for given vehicle speed, the whole range of operation of the ICE or MG O has to be explored. Therefore, the efficiency maps of the propulsors are needed (Figures 4 and 5). These are necessary also for evaluating the propulsors efficiency in each achievable working point.

3.1. Calculation of Optimal Operating Maps in Power-Split Operations

For a given vehicle speed V_{veh} and subsequent output power P_{out}, the powertrain operation can be analyzed by considering each couple of ICE speed and torque (ω_{in}, T_{in}) ranging from their minimum and maximum values within the ICE operation range of Figure 4. This leads to the calculation of an overall speed ratio matrix, having the same dimension of the ICE efficiency map, where each element is $\tau = V_{veh}/(R_w \cdot \omega_{in})$. A corresponding overall power ratio matrix containing $\eta = -P_{out}/(\omega_{in} \cdot T_{in})$ can be obtained, too. Hence, these matrices can be used to interpolate the dimensionless results of Figure 2, in order to identify the speed and power ratios of the electric machines for each combination of τ and η. Then, these ratios can be multiplied by the corresponding ω_{in} and P_{in} to assess the dimensional rotational speed of the electric machines (ω_i, ω_o) and their mechanical power ($\overline{P}_i$, $\overline{P}_o$). In this way, the operating point of both electric machines is univocally determined, as is their efficiency which enables the calculation of the net electric power flowing to or from the battery as follows:

$$P_{batt}(V_{veh}, \omega_{in}, T_{in}) = \overline{P}_i \eta_I^{-sign(\overline{P}_i)} + \overline{P}_o \eta_O^{-sign(\overline{P}_o)} \tag{8}$$

The positive sign of the power flows is shown in Figure 1.

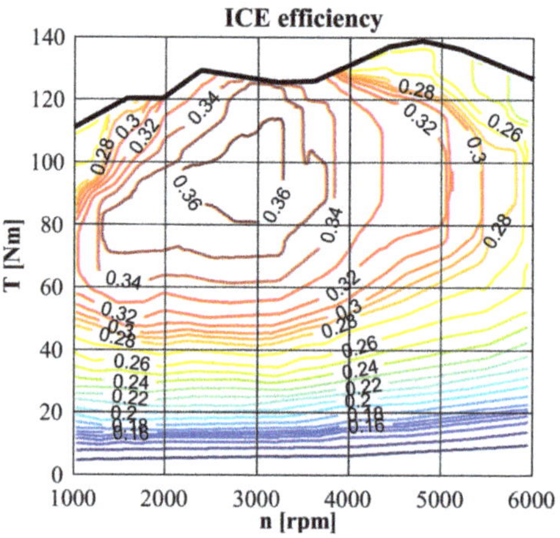

Figure 4. Efficiency map of the Chevrolet Volt ICE derived from [37].

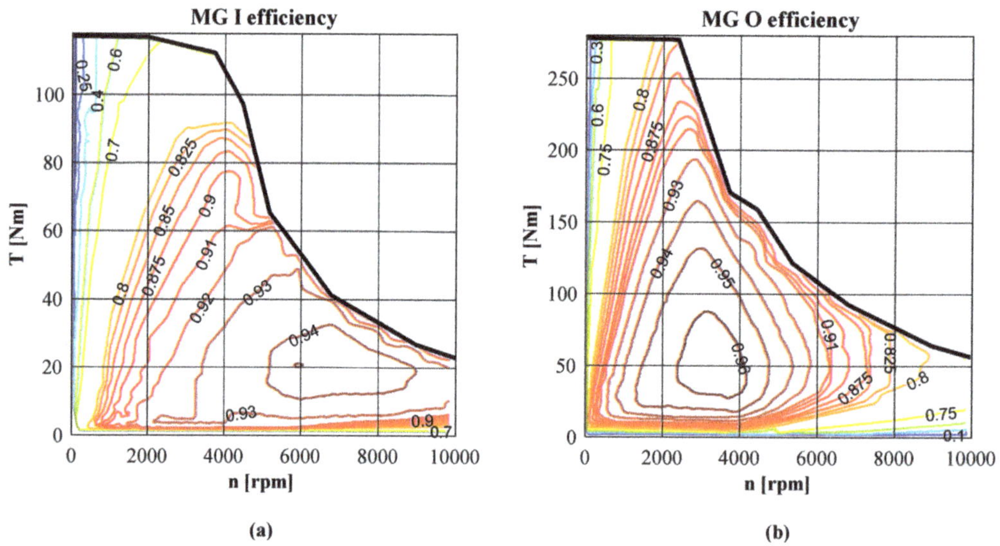

Figure 5. Efficiency maps of Chevrolet Volt electric machines derived from [38]; (a) Motor Generator I; (b) Motor Generator O. Exploited for both motoring and generating operation.

Hence, by the simple procedure herein addressed, it is possible to obtain a set of matrices containing all data describing the powertrain steady-state operation, which can be exploited as a basis of the desired energy management strategy. As an example, in this paper, a procedure for pursuing maximum global efficiency is proposed. According to the direction of the battery power, P_{batt} can be an output or input power in the powertrain. Therefore, if $P_{batt} > 0$ the battery supports the ICE in the vehicle propulsion, and the global efficiency is:

$$\eta_{gl}(V_{veh}, \omega_{in}, T_{in}) = -\frac{P_{out}}{P_{fuel} + P_{batt}} \quad (9)$$

where $P_{fuel} = P_{in}/\eta_{ICE}$ is the power provided by fuel combustion. η_{ICE} can be derived for each combination of ω_{in} and T_{in} from Figure 4. If $P_{batt} < 0$ the ICE delivers power in surplus which can be used to recharge the battery, and the global efficiency is:

$$\eta_{gl}(V_{veh}, \omega_{in}, T_{in}) = -\frac{P_{out} + P_{batt}}{P_{fuel}} \quad (10)$$

Eventually, it is possible to extract the maximum value from the matrix η_{gl} for a given vehicle speed and find the related ICE and electric machines operation resulting in the most efficient driving.

Nevertheless, the working points which violate a constructive constraint of propulsors, power converters, or batteries should not be included among the potential optimal ones. Therefore, the final operating maps do not include the functioning points whereby the ICE or electric machines operation is not included within the maps of Figures 4 and 5, or the electric powers overcome their respective maximum limits of power converters or batteries (Table 5).

Table 5. Maximum power of battery and power converters of Chevrolet Volt derived from [37,39].

Power Converter I Max. Power [kW]	Power Converter O Max. Power [kW]	Battery Pack Max. Power [kW]
48	87	120

In fact, the real constraint on P_{batt} depends on the battery state of charge (SOC):

$$P_{max\_charge}(\mathrm{SOC}) \leq P_{batt} \leq P_{max\_discharge}(\mathrm{SOC}) \quad (11)$$

P_{max_charge} and $P_{max\_discharge}$ should be evaluated instantaneously by a deeper dynamic analysis. However, a simplified approach is exploited in the following to swiftly analyze four different scenarios, summarized in Table 6:

- SOC comprised between its lower and higher thresholds (SOC = FREE);
- Battery completely discharged (SOC = 0);
- Battery completely charged (SOC = 1);
- Maintaining-charge driving (SOC = CONSTANT).

Table 6. Adjusted constraints on battery power according to the SOC.

SOC	P_{max_charge} [kW]	$P_{max\_discharge}$ [kW]
FREE	−120	120
0	−120	0
1	0	120
CONSTANT	0	0

3.2. Calculation of Optimal Operating Maps in FEV Operations

For given vehicle speed, in FEV driving the rotational speeds of the electric machines are univocally determined by Equation (1), therefore the only degree of freedom is the torque of one electric machine. As stated in Section 2, by choosing as the independent variable the torque of MG O, which can range from its minimum to maximum value corresponding to each rotational speed (Figure 5), it is possible to consider an array with different values of $p_o' = -(\omega_o \cdot T_o)/P_{out}$. This can be used to interpolate the dimensionless results of Figure 3b and find the potential operating points of MG I for given P_o and P_{out}. Then, similarly to Section 3.1, P_{batt} can be calculated by Equation (8) and the global efficiency in FEV operation is simply:

$$\eta_{gl}(V_{veh}, T_o) = -\frac{P_{out}}{P_{batt}} \tag{12}$$

In FEV operation, the battery SOC is supposed to be always sufficient to provide the demanded power.

4. Results and Discussion

The procedure described in Section 3 for calculating the optimal operating maps was implemented in MATLAB, after having carried out the dimensionless approach reported in Section 2. The following results were computed by considering a total vehicle mass equal to $m = 1750$ kg in plain ($\alpha = 0$ rad). The analyzed vehicle speed ranges from 5 to 200 km/h.

4.1. Results in Power-Split Operations

The mesh grid used in input to the script was obtained by the arrays $\omega_{in} = 1000:10:6000$ rpm and $T_{in} = 10:1:140$ Nm. The procedure described in Section 3.1 was carried out for each vehicle speed, whereby the respective optimal operating point was selected after excluding those outside the boundaries of Figures 4 and 5, Table 5, Equation (11), and Table 6. The results of Figures 6–11 show the optimal operating points resulting from the optimization procedure aimed at minimizing the powertrain power losses.

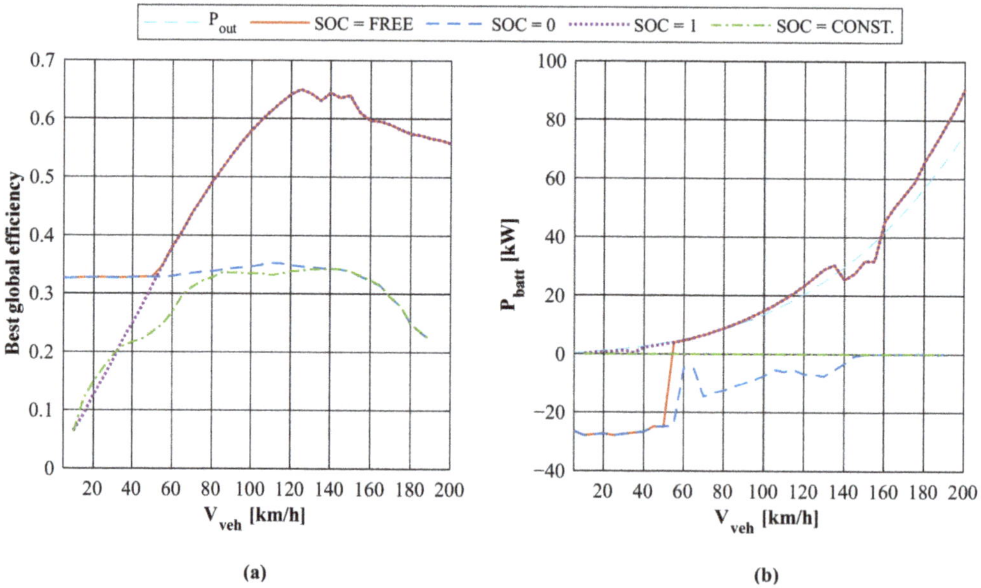

Figure 6. (**a**) Best global efficiency in power-split operation; (**b**) battery power in the optimal power-split operating points.

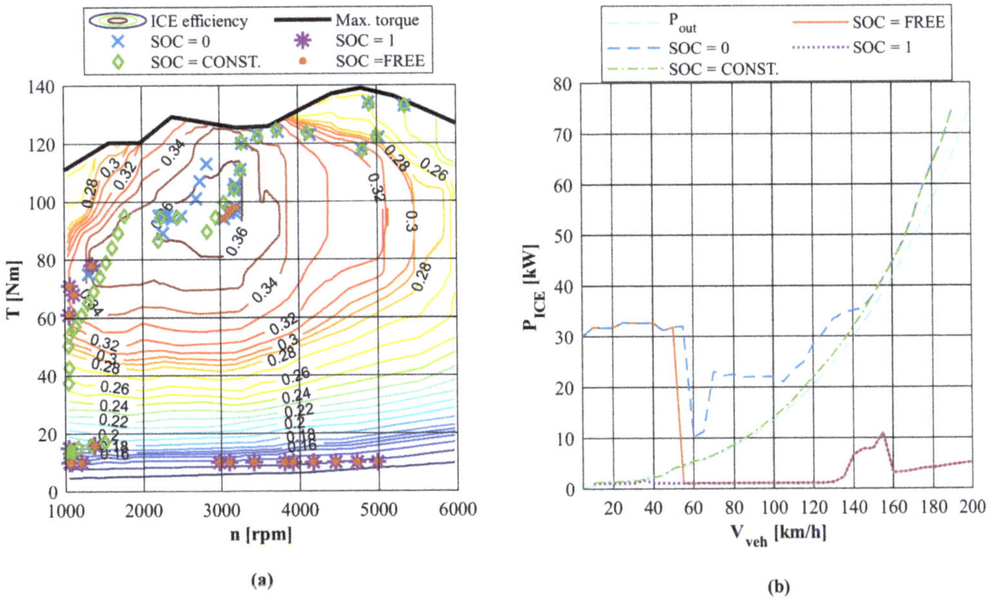

Figure 7. ICE optimal power-split operation: (**a**) ICE functioning points; (**b**) ICE power.

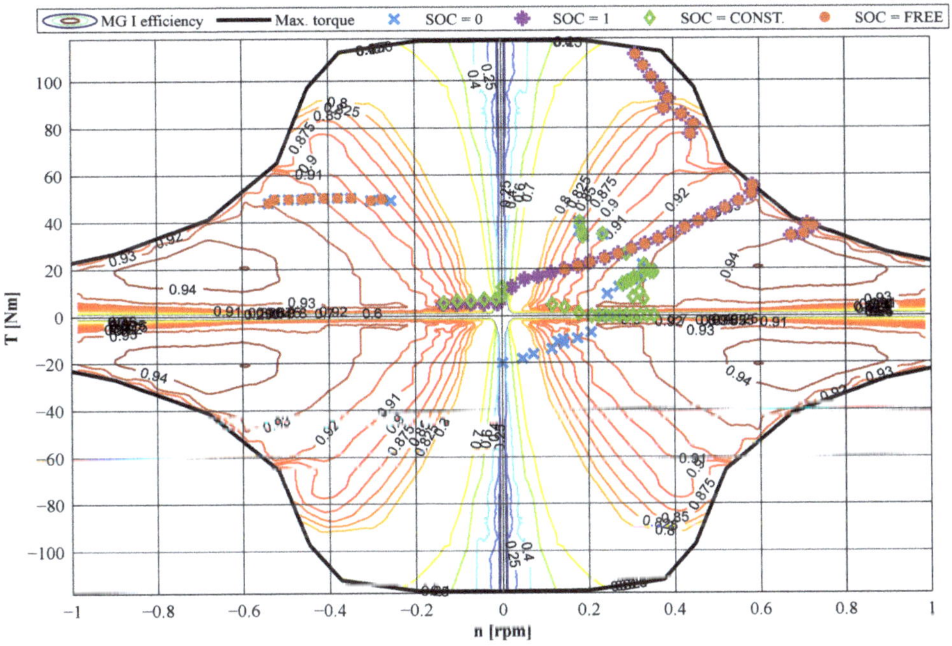

Figure 8. MG I optimal functioning points in power-split operation.

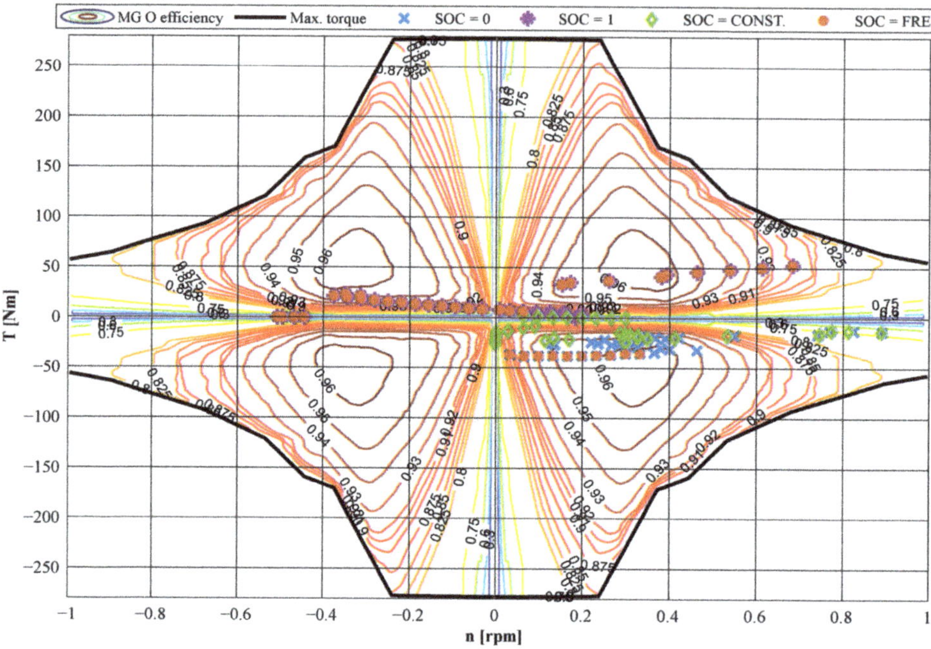

Figure 9. MG O optimal functioning points in power-split operation.

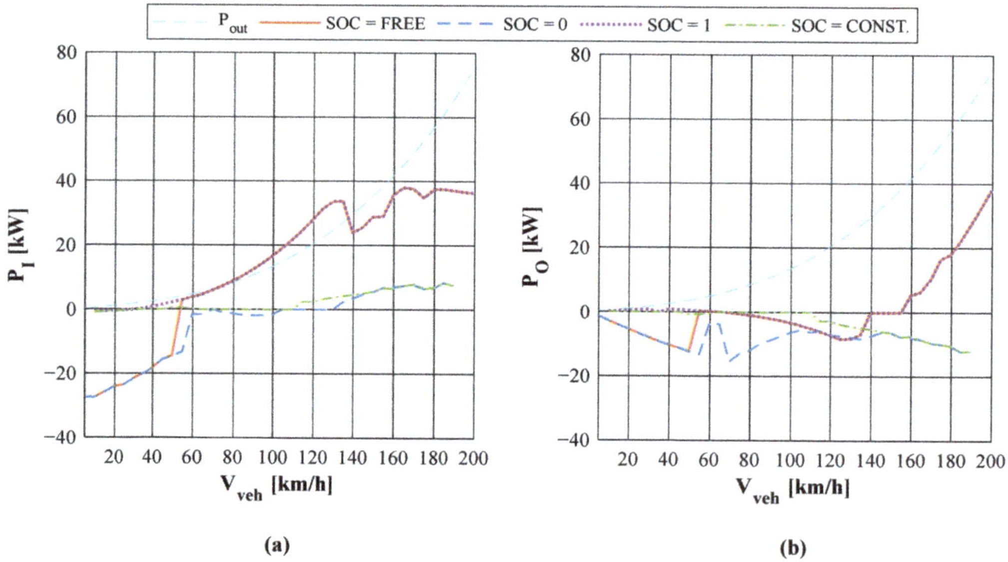

Figure 10. Electric machines optimal power in power-split operation: (**a**) MG I power; (**b**) MG O power.

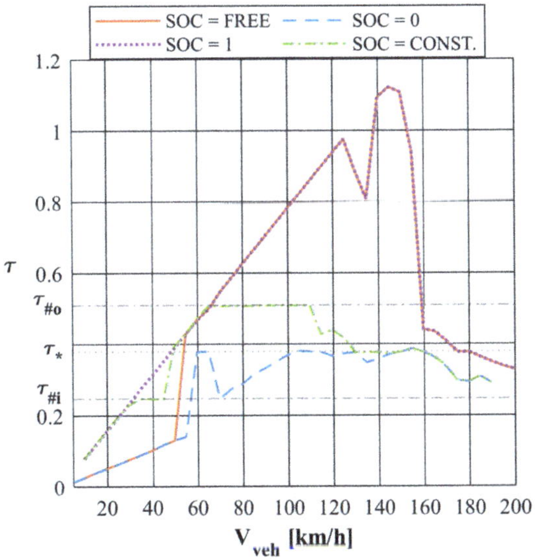

Figure 11. Optimal overall transmission ratio and mode selection in power-split operation.

Figure 6a shows that the best results are achieved for SOC = FREE since the availability of the battery both as a power source and power storage would enable the achievement of the most efficient power flows (Figure 10). Nevertheless, the maximum global efficiency is lower than 0.33 up to 50 km/h, therefore it would be more advisable to turn off the engine and drive in FEV operation (see Section 4.2). The only reason to let the ICE work at lower speeds is to recharge the battery if possible (SOC = FREE or SOC = 0). In this case, the ICE should operate in the maximum efficiency region, otherwise, it should be turned off also for higher speeds if the battery can supply power for propulsion (SOC = FREE or SOC = 1) (Figure 7). In this way, the global efficiency would be significantly enhanced (Figure 6a), since the demanded output power would be provided by the electric unit (Figure 6b), which is more efficient than the ICE.

Nevertheless, a more robust control strategy should regulate the battery power according to the instantaneous SOC, so as to ensure sufficient range. Indeed, for speeds higher than 100 km/h, the optimal powertrain operation would be achieved at the expense of the driving range. Therefore, over 100 km/h would be even more advisable to limit the power supplied by the battery and increase that provided by the ICE, even though this would reduce the global efficiency.

On the other hand, if the battery is completely discharged (SOC = 0) or a maintaining-charge driving is desired (SOC = CONSTANT), the demanded output power should be provided mainly by the engine (Figure 7b). In this case, the battery charging would be recommended between 50 and 145 km/h, while over 145 km/h keeping the SOC constant would result in greater efficiency. Nonetheless, since the ICE maximum power is 75 kW [37], the vehicle speed cannot exceed 190 km/h with SOC = 0 or SOC = CONSTANT.

Figures 8–10 show the optimal exploitation of the electric machines in power-split operation. They suggest using both MG I and MG O as generators for battery recharging up to 50 km/h. It should be the same from 50 to 130 km/h if SOC = 0, while MG I should be used as a motor and MG O as a generator if the battery can provide power (SOC = FREE or SOC = 1). In the latter case, the mechanical energy converted to electric energy by MG O is reconverted to the mechanical form by MG I. Over 155 km/h, both MG I and MG O should be used as motors for SOC = FREE or SOC = 1.

To assess the optimal mode selection, it is sufficient to analyze the optimal overall transmission ratio of Figure 11.

Figure 11 shows that for lower speeds the input-split mode should be preferred, while the compound-split mode is advisable at higher speeds. Moreover, it is worth noting that for SOC = 0 and SOC = CONSTANT (i.e., if the battery cannot provide power for propulsion) the optimal overall speed ratio at medium-high speed is the one that realizes the PGs synchronism, where the mechanical power losses are minimized (Figure 2).

4.2. Results in FEV Operations

The array used in input to the script was $T_o = -280 : 1 : 280$ Nm. The procedure described in Section 3.1 was carried out for each vehicle speed, whereby the respective optimal operating point was selected after excluding those outside the boundaries of Figure 5 and Table 5. The results of Figures 12 and 13 are related to the best powertrain operating points resulting in the highest global efficiency for each vehicle speed.

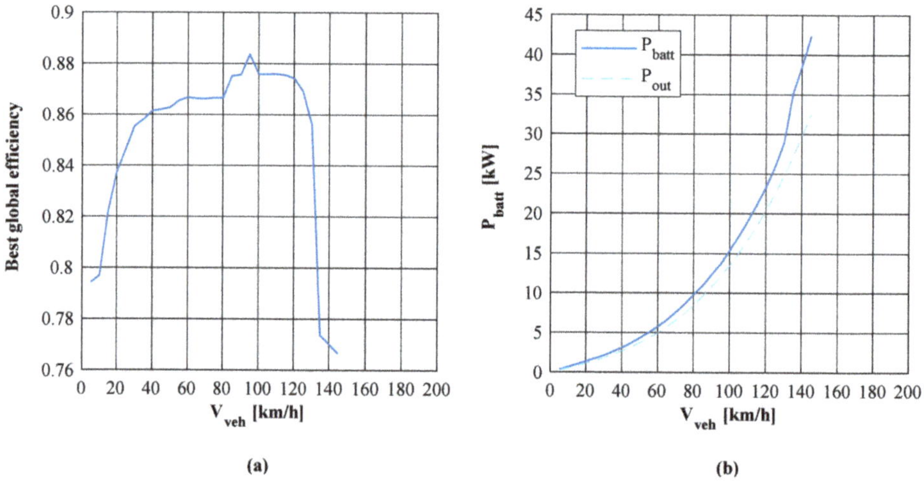

Figure 12. (**a**) Best global efficiency in FEV operation; (**b**) battery power in the optimal FEV operating points.

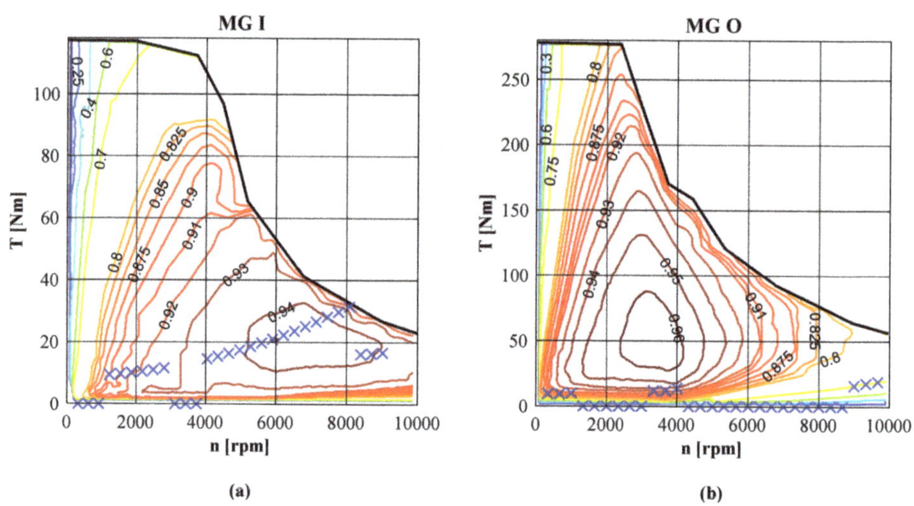

Figure 13. Electric machines optimal FEV operation, indicated by "X": (**a**) MG I functioning points; (**b**) MG O functioning points.

Figure 12a shows that the global efficiency in FEV operation is averagely higher than the one achievable in power-split operation, thus the FEV driving is specifically suggested for low–medium speed. As stated in Section 4.1, exploiting the only battery power for propelling the vehicle in steady-state driving at higher speeds would imply a reduced range, thus it should be avoided. Over 145 km/h the FEV driving cannot be realized because the required rotational speed of MG O would overcome its maximum value. Figure 13 shows that the optimal FEV operation in steady-state driving involves the exploitation of MG I except for two limited vehicle ranges from 5 to 15 km/h and from 50 to 60 km/h. From 135 to 145 it is advisable to operate with both electric machines acting as motors.

5. Conclusions

This article is an advancement of the previous contributions [33–36] addressing a unified parametric model for the analysis of PS-CVT. Herein it was applied to the Chevrolet Volt by introducing the propulsors efficiency maps to move from a dimensionless approach to a dimensional one. The final aim was to provide a complete and general tool to comprehensively analyze any power-split transmission, even multi-PG and multi-mode as Voltec is, by assessing also the mechanical power losses which are often neglected owing to the difficulty of their calculation. The FEV operation can also be swiftly analyzed.

As described in Sections 3 and 4, this method enables the investigation of the powertrain response for given vehicle speed and demanded torque. All the viable operating points are available as a set of operating maps containing the propulsors functioning points, their efficiency, the mechanical power losses in the transmission, the battery power, and the global efficiency. These data can be exploited for the development of the desired energy management strategy.

As an example, far from presenting it as an exhaustive control strategy, this paper proposes a procedure for selecting the optimal operating points which maximize the powertrain global efficiency in steady-state power-split and FEV driving. The results showed the importance of properly handling the batteries SOC to ensure the possibility of exploiting them as an energy exchanger, as well as the benefit of PGs synchronism to reduce the friction losses. Nonetheless, it seems that the MG O appears to be under-exploited in steady-state driving (Figure 9). This can be due to the fact that the surplus power of MG O could be used to provide an acceleration boost, thus it might be exploited more in dynamic operation, which was not analyzed here.

Therefore, future research could be aimed at extending the model in this very direction. Indeed, its basic relationships between PSU speed, torque, and power ratios proposed in [33,35,36] would be still valid in dynamic operation, being based on the principle of power conservation. However, the correspondence between torques developed at the PSU main ports and those provided by the propulsors would be missing because of inertial effects caused by acceleration or deceleration of the vehicle and the propulsors themselves. The extension of the model to the dynamic condition would enable a more in-depth simulation of the powertrain operation which could instantaneously consider the actual battery SOC or the power converters' efficiency. After collecting the results of this sort of simulation, they can be used to develop the most robust energy management strategies available in the literature to optimize the desired objective function, which can be even the minimization of the mechanical power losses.

Author Contributions: Methodology, M.C.; Software, A.C.; Supervision, M.C.; Writing—original draft, A.C.; Writing—review & editing, A.C. and M.C. All authors have read and agreed to the published version of the manuscript.

Funding: This research received no external funding.

Institutional Review Board Statement: Not applicable.

Informed Consent Statement: Not applicable.

Data Availability Statement: Not applicable.

Conflicts of Interest: The authors declare no conflict of interest.

Nomenclature

k_x	Fixed-gear ratio on the x-th branch
P_j or $\overline{P}_j$	Ideal or real power in the j-th branch
p_j or $\overline{p}_j$	Dimensionless ideal or real power in the j-th branch as a fraction of the input power
p'_j or $\overline{p}'_j$	Dimensionless ideal or real power in the j-th branch as a fraction of the output power
$\overline{P}_{Loss}$	Mechanical power losses
$\overline{p}_L$	Dimensionless mechanical power losses as a fraction of the input power
$\overline{p}'_L$	Dimensionless mechanical power losses as a fraction of the output power
T_j	Torque applied to the j-th shaft
η	Opposite of the overall power ratio ($\eta = -P_{out}/P_{in}$)
η_0	Fixed-carrier efficiency of a PG
η^Z	Fixed-Z apparent efficiency of a PG
η_{gl}	Powertrain global efficiency
η_I and η_O	Efficiency of the electric machines I and O
η_{ICE}	Efficiency of the internal combustion engine
$\eta_{X/x}$	Efficiency of the x-th OG
τ	Overall speed ratio
τ_j	Speed ratio of the j-th shaft
$\tau_{\#j}$	Overall speed ratio for which the j-th shaft is motionless (i.e., mechanical point or nodal ratio)
$\tau_{j\#k}$	Speed ratio of the j-th shaft when the k-th shaft is motionless (i.e., corresponding speed ratio)
τ_*	Overall speed ratio for which one or more PGs work at synchronism
$\phi^z_{x/y}$	Generalized characteristic function
Ψ	Willis' ratio of a PG
$\psi^Z_{Y/X}$	Fixed-Z speed ratio of a PG
ω_j	Rotational speed of the j-th shaft

Appendix A

The basic theory of the unified parametric model addressed in this paper considers the PSU made up of one or more three-port mechanisms (TPMs), which consists of one active PG and up to three OGs (Figure A1). X, Y, Z indicate a general way to refer to the PG branches (ring gear, sun gear, and carrier); x, y, z indicate a general way to refer to the PSU main port (in, out, i, o).

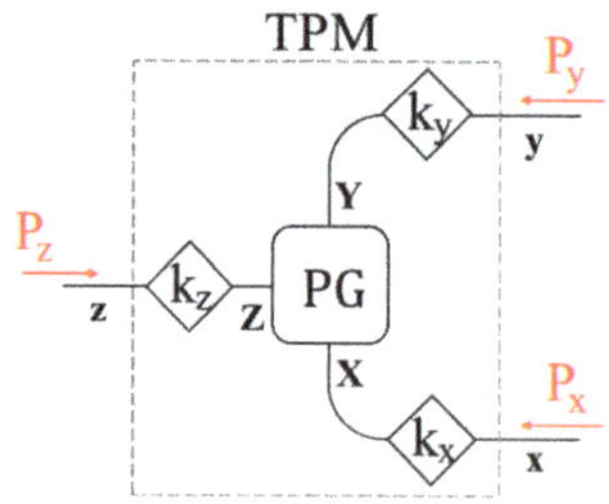

Figure A1. Scheme of a three-port mechanism with one PG (illustrated by a rounded-corner square) and three OGs (rhombi) realizing the k_j fixed-ratio. Red arrows indicate the positive direction of power flows.

The nomenclature adopted in this model distinguishes between torques and powers assessed by considering the PSU ideal or real. In the latter case, torques and powers symbols are overlined. Nonetheless, this distinction affects only the dependent variables, which are supposed to counterbalance the PSU mechanical power losses. Moreover, capital letters indicate dimensional power flows, while the dimensionless ones are indicated by lowercase letters without or with an apex, depending on whether they are normalized to the input power (P_{in}) or to the opposite of the output power ($-P_{out}$), respectively. The former normalization is exploited to address power-split operation, while the latter is used in FEV analysis.

The total mechanical power losses ($\overline{p}'_L$ in FEV operation) are the sum of the losses occurring in OGs and PGs, which can be derived as follows:

$$\overline{p}'_L\big|_{OG} = -\frac{\overline{P}_{Loss}\big|_{OG}}{P_{out}} \approx -\left|(1-\eta_{X/x})\, p'_x\right| \tag{A1}$$

$$\overline{p}'_L\big|_{PG} = -\frac{\overline{P}_{Loss}\big|_{PG}}{P_{out}} \approx -\left|\left(1-\eta^Z\right)\left(\frac{\phi^z_{x/y}-\psi^Z_{X/Y}}{1-\psi^Z_{X/Y}}\right)p'_x\right| \tag{A2}$$

In Equation (A1) $\eta_{X/x}$ is the OG efficiency, while in Equations (A1) and (A2) $p'_x = -P_x/P_{out}$ is the normalized power ratio of the x-th shaft. In Equation (A2) η^Z is the efficiency of the PG evaluated when its branch Z is still, while $\psi^Z_{X/Y}$ is the speed ratio between branches X and Y when Z is still. Therefore, both η^Z and $\psi^Z_{X/Y}$ are constant parameters depending on the PG Willis' ratio Ψ and its basic fixed-carrier efficiency η_0 [36]. $\phi^z_{x/y}$ is the characteristic function, crucial tool for both analysis and design purpose [33,36], which is a function of τ ruled by nodal ratios:

$$\phi^z_{x/y} = \frac{\tau_x/\tau_y\big|_{\tau_z=0}}{\tau_x/\tau_y} = \frac{\tau_{\#z}-\tau_{\#x}}{\tau_{\#z}-\tau_{\#y}}\cdot\frac{\tau-\tau_{\#y}}{\tau-\tau_{\#x}} \tag{A3}$$

It is worth noting that in FEV operation $\tau \to \infty$; this can be modeled in numeric software by considering a value of τ high enough (e.g., 10^5). Moreover, the characteristic functions rule the ideal power ratio between two TPM shafts:

$$\phi^z_{x/y} = -\frac{P_y}{P_x} = -\frac{p'_y}{p'_x} \tag{A4}$$

References

1. Guzzella, L.; Sciarretta, A. *Vehicle Propulsion Systems: Introduction to Modeling and Optimization*; Springer: Berlin/Heidelberg, Germany, 2013; ISBN 9783642359132.
2. Mi, C.; Abul Masrur, M. *Hybrid Electric Vehicles, Principles and Applications with Practical Perspectives*, 2nd ed.; Wiley: Hoboken, NJ, USA, 2018; ISBN 9780470747735.
3. Krithika, V.; Subramani, C. A comprehensive review on choice of hybrid vehicles and power converters, control strategies for hybrid electric vehicles. *Int. J. Energy Res.* **2018**, *42*, 1789–1812. [CrossRef]
4. Zhuang, W.; Li, S.; Zhang, X.; Kum, D.; Song, Z.; Yin, G.; Ju, F. A survey of powertrain configuration studies on hybrid electric vehicles. *Appl. Energy* **2020**, *262*, 114553. [CrossRef]
5. Singh, K.V.; Bansal, H.O.; Singh, D. A comprehensive review on hybrid electric vehicles: Architectures and components. *J. Mod. Transp.* **2019**, *27*, 77–107. [CrossRef]
6. Zeng, X.; Wang, J. *Analysis and Design of the Power-Split Device for Hybrid Systems*, 1st ed.; Springer: New York, NY, USA, 2017; ISBN 9789811042720.
7. Chen, H.; Li, L.; Lange, A.; Küçükay, F. Innovative Dedicated Hybrid Transmission Concepts in the Next Generation of Hybrid Powertrains. *SAE Int. J. Altern. Powertrains* **2019**, *8*, 75–88. [CrossRef]
8. Sieg, C.; Küçükay, F. Benchmarking of Dedicated Hybrid Transmissions. *Vehicles* **2020**, *2*, 100–125. [CrossRef]
9. Kim, N.; Kwon, J.; Rousseau, A. Trade-off between multi-mode powertrain complexity and fuel consumption. In Proceedings of the 25th World Battery, Hybrid and Fuel Cell Electric Vehicle Symposium & Exhibition, Shenzhen, China, 5–9 November 2010.
10. Zhuang, W.; Zhang, X.; Ding, Y.; Wang, L.; Hu, X. Comparison of multi-mode hybrid powertrains with multiple planetary gears. *Appl. Energy* **2016**, *178*, 624–632. [CrossRef]
11. De Pinto, S.; Mantriota, G. Power Flows in Compound Transmissions for Hybrid Vehicles. *Machines* **2019**, *7*, 19. [CrossRef]
12. Bottiglione, F.; Mantriota, G. Power Flows and Efficiency of Output Compound e-CVT. *Int. J. Veh. Technol.* **2015**, *2015*, 136437. [CrossRef]

13. Gupta, A.K.; Ramanarayanan, C.P. Analysis of circulating power within hybrid electric vehicle transmissions. *Mech. Mach. Theory* **2013**, *64*, 131–143. [CrossRef]
14. Panday, A.; Bansal, H.O. A review of optimal energy management strategies for hybrid electric vehicle. *Int. J. Veh. Technol.* **2014**, *2014*, 160510. [CrossRef]
15. Zhang, F.; Wang, L.; Coskun, S.; Pang, H.; Cui, Y.; Xi, J. Energy management strategies for hybrid electric vehicles: Review, classification, comparison, and outlook. *Energies* **2020**, *13*, 3352. [CrossRef]
16. Zhang, P.; Yan, F.; Du, C. A comprehensive analysis of energy management strategies for hybrid electric vehicles based on bibliometrics. *Renew. Sustain. Energy Rev.* **2015**, *48*, 88–104. [CrossRef]
17. Enang, W.; Bannister, C. Modelling and control of hybrid electric vehicles (A comprehensive review). *Renew. Sustain. Energy Rev.* **2017**, *74*, 1210–1239. [CrossRef]
18. Pennestrì, E.; Mariti, L.; Valentini, P.P.; Mucino, V.H. Efficiency evaluation of gearboxes for parallel hybrid vehicles: Theory and applications. *Mech. Mach. Theory* **2012**, *49*, 157–176. [CrossRef]
19. Esmail, E.L.; Pennestrì, E.; Cirelli, M. Power-Flow and Mechanical Efficiency Computation in Two-Degrees-of-Freedom Planetary Gear Units: New Compact Formulas. *Appl. Sci.* **2021**, *11*, 5991. [CrossRef]
20. Cammalleri, M. Efficiency of split-way CVT's. A simplified model. *SAE Trans.* **2007**, *116*. [CrossRef]
21. Yang, F.; Feng, J.; Zhang, H. Power flow and efficiency analysis of multi-flow planetary gear trains. *Mech. Mach. Theory* **2015**, *92*, 86–99. [CrossRef]
22. De Carlo, M.; Mantriota, G. Electric vehicles with two motors combined via planetary gear train. *Mech. Mach. Theory* **2020**, *148*, 103789. [CrossRef]
23. Mantriota, G.; Reina, G. Dual-Motor Planetary Transmission to Improve Efficiency in Electric Vehicles. *Machines* **2021**, *9*, 58. [CrossRef]
24. Kang, J.; Choi, W.; Kim, H. Development of a control strategy based on the transmission efficiency with mechanical loss for a dual mode power split-type hybrid electric vehicle. *Int. J. Automot. Technol.* **2012**, *13*, 825–833. [CrossRef]
25. Anselma, P.G.; Biswas, A.; Belingardi, G.; Emadi, A. Rapid assessment of the fuel economy capability of parallel and series-parallel hybrid electric vehicles. *Appl. Energy* **2020**, *275*, 115319. [CrossRef]
26. Kim, N.D.; Kim, J.M.; Kim, H.S. Control strategy for a dual-mode electromechanical, infinitely variable transmission for hybrid electric vehicles. *Proc. Inst. Mech. Eng. Part D J. Automob. Eng.* **2008**, *222*, 1587–1601. [CrossRef]
27. Park, J.; Park, J.-H. Development of equivalent fuel consumption minimization strategy for hybrid electric vehicles. *Int. J. Automot. Technol.* **2012**, *13*, 835–843. [CrossRef]
28. Lee, W.; Park, J.; Kim, N. Analysis of Transmission Efficiency of a Plug-In Hybrid Vehicle Based on Operating Modes. *Int. J. Precis. Eng. Manuf.-Green Technol.* **2021**, *8*, 165–175. [CrossRef]
29. Kim, N.; Choi, S.; Jeong, J.; Vijayagopal, R.; Stutenberg, K.; Rousseau, A. Vehicle level control analysis for Voltec powertrain. *World Electr. Veh. J.* **2018**, *9*, 29. [CrossRef]
30. Zhuang, W.; Zhang, X.; Li, D.; Wang, L.; Yin, G. Mode shift map design and integrated energy management control of a multi-mode hybrid electric vehicle. *Appl. Energy* **2017**, *204*, 476–488. [CrossRef]
31. Park, J.Y.; Park, Y.K.; Park, J.H. Optimal power distribution strategy for series-parallel hybrid electric vehicles. *Proc. Inst. Mech. Eng. Part D J. Automob. Eng.* **2008**, *222*, 989–1000. [CrossRef]
32. Ahn, K.; Cho, S.; Cha, S.W.; Lee, J.M. Engine operation for the planetary gear hybrid powertrain. *Proc. Inst. Mech. Eng. Part D J. Automob. Eng.* **2006**, *220*, 1727–1735. [CrossRef]
33. Cammalleri, M.; Rotella, D. Functional design of power-split CVTs: An uncoupled hierarchical optimized model. *Mech. Mach. Theory* **2017**, *116*, 294–309. [CrossRef]
34. Rotella, D.; Cammalleri, M. Direct analysis of power-split CVTs: A unified method. *Mech. Mach. Theory* **2018**, *121*, 116–127. [CrossRef]
35. Rotella, D.; Cammalleri, M. Power losses in power-split CVTs: A fast black-box approximate method. *Mech. Mach. Theory* **2018**, *128*, 528–543. [CrossRef]
36. Cammalleri, M.; Castellano, A. Analysis of hybrid vehicle transmissions with any number of modes and planetary gearing: Kinematics, power flows, mechanical power losses. *Mech. Mach. Theory* **2021**, *162*, 104350. [CrossRef]
37. Conlon, B.M.; Blohm, T.; Harpster, M.; Holmes, A.; Palardy, M.; Tarnowsky, S.; Zhou, L. The Next Generation "voltec" Extended Range EV Propulsion System. *SAE Int. J. Altern. Powertrains* **2015**, *4*, 248–259. [CrossRef]
38. Jurkovic, S.; Rahman, K.; Patel, N.; Savagian, P. Next Generation Voltec Electric Machines; Design and Optimization for Performance and Rare-Earth Mitigation. *SAE Int. J. Altern. Powertrains* **2015**, *4*, 336–342. [CrossRef]
39. Anwar, M.; Hayes, M.; Tata, A.; Teimorzadeh, M.; Achatz, T. Power dense and robust traction power inverter for the second-generation chevrolet volt extended-range EV. *SAE Int. J. Altern. Powertrains* **2015**, *4*, 145–152. [CrossRef]

applied sciences

MDPI

Article

Power-Flow and Mechanical Efficiency Computation in Two-Degrees-of-Freedom Planetary Gear Units: New Compact Formulas

Essam Lauibi Esmail [1,†], Ettore Pennestrì [2,*,†] and Marco Cirelli [3]

1 Department of Mechanical Engineering, University of Al-Qadisiyah, Diwaniya 0964, Iraq; essam.esmail@qu.edu.iq
2 Department of Enterprise Engineering, University of Rome Tor Vergata, 00133 Rome, Italy
3 Deparment of Mechanical Engineering, University Niccolò Cusano, 00166 Rome, Italy; marco.cirelli@unicusano.it
* Correspondence: pennestri@mec.uniroma2.it; Tel.: +39-06-7259-7138
† These authors contributed equally to this work.

Abstract: The mechanical efficiency is a computed value for comparing the performance of the multi degrees-of-freedom geared transmissions of hybrid vehicles. Most of the current methods for estimating gear trains mechanical efficiency require the decomposition of gear transmissions in basic structural elements or planetary gear units (PGU). These are two degrees-of-freedom components whose mechanical efficiency has a deep influence on the overall device. The authors (E.L.E., E.P.) already evidenced that, under certain kinematic conditions, the classic Radzimovsky's formulas, widely accepted for computing the mechanical efficiency of PGUs, are not adequate. In this paper, more general and reliable formulas for computing the mechanical efficiency are deduced. The proposed formulas herein, exploiting the concept of potential or virtual power, evidence the dependency between kinematics and efficiency. A numerical example compares our results with previous work on the subject.

Keywords: planetary gear trains; mechanical efficiency; power-flow; hybrid vehicles

Citation: Esmail, E.L.; Pennestrì, E.; Cirelli, M. Power-Flow and Mechanical Efficiency Computation in Two-Degrees-of-Freedom Planetary Gear Units: New Compact Formulas . *Appl. Sci.* **2021**, *11*, 5991. https://doi.org/10.3390/app11135991

Academic Editors: Edris Pouresmaeil, Marco Cammalleri and Vincenzo Di Dio

Received: 31 May 2021
Accepted: 24 June 2021
Published: 28 June 2021

1. Introduction

The multi degrees-of-freedom gear drive is a key component in hybrid vehicles. Its capability of managing power flows from different sources is of paramount importance for energy saving. For these reasons, the mechanical efficiency of planetary multi degrees-of-freedom planetary gear trains is a topic that received many contributions in recent times (e.g., [1–14]).

The analysis carried out herein offers a reliable approach toward a synthetic description of the relationship between kinematics conditions and gear train mechanical efficiency.

In this paper, the kinematic conditions governing the relationship between the algebraic signs of actual and virtual or potential powers are elucidated for the first time. Such an analysis has a significant influence on the casting of physically reliable and consistent formulas for the computation of mechanical efficiency in a two dof PGU (Planetary Gear Unit). Subsequently, two general formulas for mechanical efficiency computation in a two dof PGU are proposed. The differential unit is the building block of more complex planetary gear trains. Thus, the understanding of the two dof PGU basic mechanics influences the design of the entire planetary gear drive [15].

The approach adopted herein is based on the concept of *virtual* or *potential* power, at the base of thoughtful contributions on the mechanical efficiency of planetary gear trains [5,16–22]. The *virtual power*, a term coined by Chen and Angeles [17], or *potential power*, coined by Esmail [5], is the power measured by an observer on any arbitrary moving frame [23,24]. However, the use of kinematic inversion for torque and efficiency analysis on gear trains dates back to Macmillan [16]. In particular, he recognized that the torques acting on the PGU links and power losses are independent of the observer's motion. The following quote is taken from [16]:

> *...our analysis is based upon an important principle relating to torques and the power lost in friction; this is the fact that magnitudes of the torques acting upon the various members of the gear are quite independent of the motion of the observer who measures them. In addition, the power lost, being determined solely by the internal torques and the relative motions of the wheels within the gear, is also independent of the observer's motion.*

Although Macmillan's analysis was limited to one dof PGU, the importance of kinematic inversion in the gear trains mechanical efficiency analysis is well represented.

The present paper was stimulated by the discovery that, under certain kinematic conditions, the classic Radzimovsky formulas [15,25–27] for the computation of mechanical efficiency in a two dof gear unit are not valid anymore [28]. This situation may lead to an estimation of the overall mechanical efficiency flawed by errors.

As discussed by Pennestrì & Valentini [15], the Radzimovsky formulas consider the two dof epicyclic gear train as two single dof devices in parallel and crossed by a power-flow. These formulas are deduced after application of the general mechanical efficiency formula of a mechanical system composed of two parallel devices with a given mechanical efficiency and power-flow direction. However, with the introduction of the concept of *virtual* or *potential* power, the mechanical efficiency of an epicyclic gear train is reduced to that of an ordinary gear train. For this purpose, a kinematic inversion is introduced such that the observed motion of the gear carrier is canceled.

The orientation of power flow observed under such kinematic inversion may not be the same as one of epicyclic arrangement. This situation requires special attention and the algebraic expression of the overall mechanical efficiency should be set in a consistent manner. The new proposed formulas herein are of a general nature and do not suffer any of the mentioned pitfalls.

In this paper, the concept of virtual or potential power was applied to deduce two new general mechanical efficiency formulas that cover all the working modes of a two dof PGU:

- Two driving members (i.e., those with positive powers) and one driven member (i.e., the one with negative power);
- Two driven members and one driving member.

Our approach is based only on the application of mechanics *first principles*, namely, Willis' formula, torque equilibrium, kinematic inversion, power balance and definition of mechanical efficiency. As a byproduct, general formulas of the power-flow ratios, for the case of a gear unit without power losses, are also deduced.

For the mechanical efficiency analysis, members with constant velocities and meshing losses only are considered. A constant mechanical efficiency is assumed for the unit working as an ordinary gear train. Moreover, the algebraic sign of the power-flow is not altered by friction.

On the basis of the previous hypotheses, two new compact formulas for the mechanical efficiency analysis of two dof PGU are deduced herein. The formulas can be adapted to any power flow arrangement within the PGU.

2. Power-Flow Ratios in a PGU

The PGU (see Figure 1) is composed of two mating gears and the gear carrier. Let us denote by x, y and z the three links of the PGU.

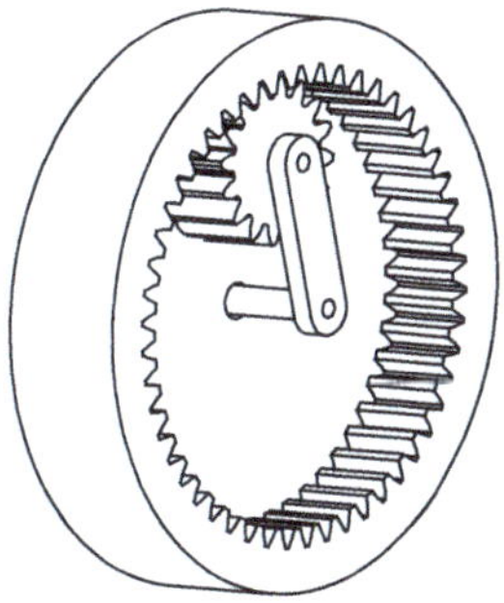

Figure 1. Example of planetary gear unit (PGU) with internally meshing gears.

Angular speeds, torques and powers in a GPU are governed, respectively, by the well-known analytical conditions from Willis' equation, torque equilibrium and power balance:

$$\omega_z + \tau_1\omega_x + \tau_2\omega_y = 0 \tag{1a}$$

$$T_x + T_y + T_z = 0 \tag{1b}$$

$$T_x\omega_x + T_y\omega_y + T_z\omega_z = 0 \tag{1c}$$

where τ_1 and τ_2 depend on the number of teeth ratio.

The readers will readily recognize (1a) as the Willis' equation. In fact, setting z and x as the subscripts denoting the meshing gears and y the gear carrier, then

$$\frac{\omega_z - \omega_y}{\omega_x - \omega_y} = N_{x,z} = \pm\frac{\text{No. of Teeth of gear } x}{\text{No. of Teeth of gear } z} \tag{2}$$

The comparison of (1a) with (2), yields $\tau_1 = -N_{x,z}$ and $\tau_2 = N_{x,z} - 1$. Obviously, other subscript combinations are possible, with corresponding different analytical expressions of τ_1 and τ_2. The choice x, y and z for the initial indistinct labeling all the moving bodies composing the PGU is justified by the advantage of condensing all possible cases for mechanical efficiency computation in two formulas only. In other words, what is lost in apparent initial abstraction is later gained in the results' generality.

After some algebraic manipulation of (1) one obtains the following:

$$\frac{P_x}{P_y} = -\frac{(1+\tau_2) + \tau_1\dfrac{\omega_x}{\omega_y}}{(1+\tau_1) + \tau_2\dfrac{\omega_y}{\omega_x}} \tag{3}$$

$$\frac{P_y}{P_z} = -\frac{(1+\tau_1) + \tau_2\dfrac{\omega_y}{\omega_z}}{(\tau_1+\tau_2) + \dfrac{\omega_z}{\omega_y}} \tag{4}$$

where $P_x = T_x\omega_x$, $P_y = T_y\omega_y$ and $P_z = T_z\omega_z$ are the powers sustained by links x, y and z, respectively. The power ratios deduced by Pennestrì & Freudenstein [29] are particular cases of the previous equations.

Under ideal conditions, in a PGU, the sign of powers is established prescribing the following: two angular speeds and one torque (kinematically driven PGU) or, alternatively, one angular speed and two torques (torque driven PGU) [30].

In a two dof PGU we distinguish the following working modes:

- Two driving links (namely x and y, $P_x > 0$ and $P_y > 0$) and one driving link (namely z, $P_z < 0$);

- Two driving links (namely x and y, $P_x < 0$ and $P_y < 0$) and one driving link (namely z, $P_z > 0$).

In the current analysis, it is assumed that the presence of meshing losses does not alter the algebraic sign of actual power flows.

3. The Algebraic Sign of Virtual Power Flows

In a previous paper [28], the authors (E.L.E. and E.P.) demonstrated that, under certain kinematic conditions, the Radzimovsky's formulas do not hold. An analysis of power-flows algebraic signs is required to explain the limits for the Radzimovsky's formulas and offer a more general alternative.

The following analysis will ascertain the algebraic sign correlation between the actual powers $P_x = T_x\omega_x$ and $P_y = T_y\omega_y$ and the *virtual* [17] or *potential* [5] powers, such as $P_x^y = T_x(\omega_x - \omega_y)$ and $P_y^x = T_y(\omega_y - \omega_x)$. For instance, the *potential power* corresponds to the power which would be transmitted by the gear unit, operating in a rotating reference frame at which link j appears relatively fixed and at relative angular velocity $\omega_u - \omega_j$, for link u. Therefore, the carrier potential power is measured under a kinematic inversion, making *fixed* the planet-carrier in the observer frame of reference. The potential power is the most important and basic principle of the mechanical efficiency analysis of differential devices).

Powers P_x and P_x^y have the same algebraic sign only if the following is true:

$$\frac{P_x^y}{P_x} = \frac{\omega_x - \omega_y}{\omega_x} > 0 \tag{5}$$

or

$$\frac{\omega_y}{\omega_x} < 1 \tag{6}$$

Similarly, P_y and P_y^x have the same algebraic sign only if the following are true:

$$\frac{P_y^x}{P_y} = \frac{\omega_y - \omega_x}{\omega_y} > 0 \tag{7}$$

or

$$\frac{\omega_x}{\omega_y} < 1 \tag{8}$$

There are two ranges consistent with inequality (6):

1. $\frac{\omega_y}{\omega_x} < 0$
2. $0 < \frac{\omega_y}{\omega_x} < 1$

Moreover, we observe the following:

- If $\frac{\omega_y}{\omega_x} < 0$, then $\frac{\omega_x}{\omega_y}$ is also negative.
- If $0 < \frac{\omega_y}{\omega_x} < 1$, then $\frac{\omega_x}{\omega_y}$ must be greater than one.

Case 1: ω_x and ω_y have opposite algebraic signs.

In this case, Equations (6) and (8) are both simultaneously valid. P_x^y and P_x, P_y^x and P_y have the same algebraic sign for any working mode, as depicted in the PGU block schemes shown in Figure 2 (1st working mode).

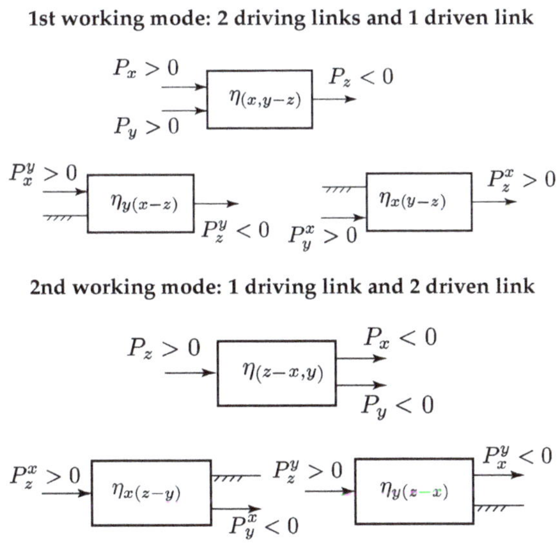

Figure 2. Virtual power flow directions Kinematic condition: $\text{sign}\left(\frac{\omega_x}{\omega_y}\right) = -1$.

Case 2: ω_x and ω_y have the same algebraic sign, or $\text{sign}\left(\frac{\omega_y}{\omega_x}\right) = +1$.

Under this hypothesis, Equations (6) and (8) cannot hold simultaneously.

Subcase 2a: When $\frac{\omega_y}{\omega_x} < 1$, then P_x^y and P_x have the same algebraic sign. The described situation is depicted in Figure 3, 1st working mode.

Subcase 2b: When $\frac{\omega_x}{\omega_y} < 1$, then P_y^x and P_y will have the same algebraic sign. The described situation is depicted in Figure 4, 1st working mode.

The previous analysis can be summarized as follows:

1. When ω_x and ω_y do not have the same algebraic sign, the power ratios $\frac{P_x^y}{P_x}$ and $\frac{P_y^x}{P_y}$ will have simultaneously the same algebraic sign. For all the links, the direction of the virtual power-flow is the same as the actual power flow (see Figure 2).
2. When ω_x and ω_y have the same algebraic sign, power ratios $\frac{P_x^y}{P_x}$ and $\frac{P_y^x}{P_y}$ cannot have simultaneously the same algebraic sign. In particular, the condition $\frac{P_x^y}{P_x} > 0$ is fulfilled when $\frac{\omega_y}{\omega_x} > 1$ (See Figure 3). Conversely, the condition $\frac{P_y^x}{P_y} > 0$ is fulfilled when $\frac{\omega_x}{\omega_y} > 1$.

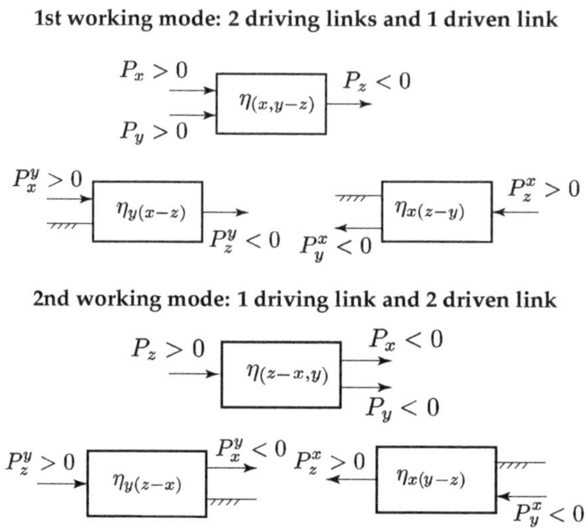

Figure 3. Virtual power-flow directions for subcase 2a. Kinematic condition: $\text{sign}\left(\frac{\omega_y}{\omega_x}\right) = +1$ and $\frac{\omega_y}{\omega_x} < 1$.

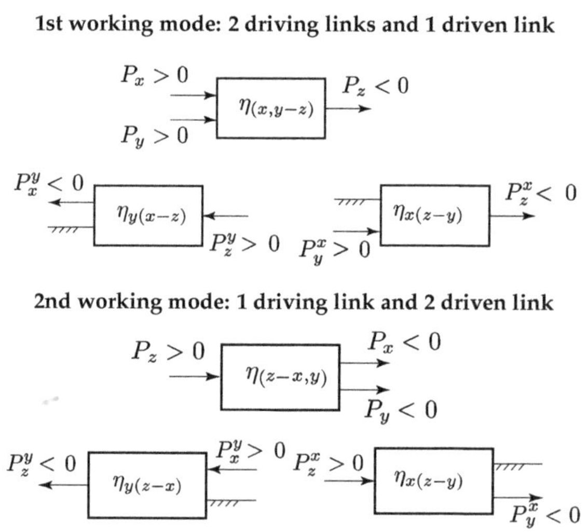

Figure 4. Virtual power flow directions for subcase 2b. Kinematic condition: $\text{sign}\left(\frac{\omega_y}{\omega_x}\right) = +1$ and $\frac{\omega_x}{\omega_y} < 1$.

4. The Modified Radzimovsky Formulas

4.1. Case x and y as Driving Links and z Driven Link

Let us consider the case of a two dof gear train with x and y as driving links (i.e., $P_x = T_x\omega_x > 0$, $P_y = T_y\omega_y > 0$ and $P_z = T_z\omega_z < 0$).

Case 1: $\text{sign}\left(\frac{\omega_x}{\omega_y}\right) = -1$ (see Figure 2, 1st working mode)

Under kinematic inversions that fix links x and y, respectively, assuming without loss of generality $P_x^y > 0$, the following relations hold (see 1st working mode of Figure 2):

$$T_x \omega_{xy} \eta_{y(x-z)} + T_z \omega_{zy} = 0 \tag{9a}$$

$$T_y \omega_{yx} \eta_{x(y-z)} + T_z \omega_{zx} = 0 \tag{9b}$$

The overall power balance condition is as follows:

$$\eta_{(x,y-z)} = \frac{|T_z \omega_z|}{T_x \omega_x + T_y \omega_y} \tag{9c}$$

Solving the system composed of the Equations (9), one obtains the following:

$$\eta_{(x,y-z)} = \left| \frac{\omega_z \omega_{yx} \eta_{y(x-z)}}{\omega_x \omega_{zy} - \dfrac{\eta_{y(x-z)}}{\eta_{x(y-z)}} \omega_y \omega_{zx}} \right| \tag{10}$$

Subcase 2a: $\dfrac{\omega_y}{\omega_x} < 1$ or $\mathrm{sgn}\left(\dfrac{\omega_x}{\omega_y}\right) = +1$ (see Figure 3, 1st working mode)

Under kinematic inversions that fix links x and y, respectively, assuming without loss of generality $P_x^y > 0$, the following relations hold:

$$T_x \omega_{xy} \eta_{y(x-z)} + T_z \omega_{zy} = 0 \tag{11a}$$

$$T_y \omega_{yx} + T_z \omega_{zx} \eta_{x(z-y)} = 0 \tag{11b}$$

Solving the system composed of the Equations (9c) and (11), one obtains the following:

$$\eta_{(x,y-z)} = \left| \frac{\omega_z \omega_{yx} \eta_{y(x-z)}}{\omega_x \omega_{zy} - \eta_{x(z-y)} \eta_{y(x-z)} \omega_y \omega_{zx}} \right| \tag{12}$$

The analytical Expression (10) can be combined into the following formula:

$$\boxed{\eta_{(x,y-z)} = \left| \frac{\omega_z \omega_{yx} \eta_{y(x-z)}}{\omega_x \omega_{zy} - A \omega_y \omega_{zx}} \right|} \tag{13}$$

with

$$A = \begin{cases} \dfrac{\eta_{y(x-z)}}{\eta_{x(y-z)}} & \text{when } \mathrm{sign}\left(\dfrac{\omega_x}{\omega_y}\right) = -1 \\ \eta_{x(z-y)} \eta_{y(x-z)} & \text{when } \left(\dfrac{\omega_y}{\omega_x}\right) < 1 \text{ and } \mathrm{sign}\left(\dfrac{\omega_x}{\omega_y}\right) = +1 \end{cases} \tag{14}$$

The proposed formula is more general than the one proposed by Radzimovsky (see Equation (11) of [26]).

When $\omega_x = \omega_y = \omega_z$, the Equations (9b) and (11) are identically satisfied and Formula (13) is not valid. However, under these kinematic conditions, there is no meshing power loss. Under our hypotheses, the mechanical efficiency is equal to one.

4.2. Case x and y as Driven Links and z Driving Link

Let us consider the case of a two dof gear train with x and y as driven links (i.e., $P_x = T_x \omega_x < 0$, $P_y = T_y \omega_y < 0$ and $P_z = T_z \omega_z > 0$), as shown in Figure 2, 2nd working mode. In this case the overall power balance equation is the following:

$$\eta_{(z-x,y)} = \frac{|T_x \omega_x + T_y \omega_y|}{T_z \omega_z} \tag{15}$$

Case 1: $\text{sign}\left(\frac{\omega_x}{\omega_y}\right) = -1$ (see Figure 2, 2nd working mode)

The following relations hold:

$$T_z\omega_{zy}\eta_{y(z-x)} + T_x\omega_{xy} = 0 \tag{16a}$$

$$T_z\omega_{zx}\eta_{x(z-y)} + T_y\omega_{yx} = 0 \tag{16b}$$

Solving the system composed of the Equations (15) and (16), one obtains the following:

$$\eta_{(z-x,y)} = \left|\frac{\eta_{x(z-y)}\omega_y\omega_{zx} - \eta_{y(z-x)}\omega_x\omega_{zy}}{\omega_{yx}\omega_z}\right| \tag{17}$$

Subcase 2a: $\frac{\omega_y}{\omega_x} < 1$ and $\text{sgn}\left(\frac{\omega_x}{\omega_y}\right) = +1$ (see Figure 3, 2nd working mode)

The following relations hold:

$$T_z\omega_{zy}\eta_{y(z-x)} + T_x\omega_{xy} = 0 \tag{18a}$$

$$T_z\omega_{zx} + T_y\omega_{yx}\eta_{x(y-z)} = 0 \tag{18b}$$

Solving the system composed of the Equations (15) and (18), one obtains the following:

$$\eta_{(z-x,y)} = \left|\frac{\omega_y\omega_{zx} - \eta_{x(y-z)}\eta_{y(z-x)}\omega_x\omega_{zy}}{\eta_{x(y-z)}\omega_{yx}\omega_z}\right| \tag{19}$$

The analytical Expressions (18) and (19) can be combined into the following formula:

$$\boxed{\eta_{(z-x,y)} = \left|\frac{B\omega_y\omega_{zx} - \eta_{y-(z-x)}\omega_x\omega_{zy}}{\omega_{yx}\omega_z}\right|} \tag{20}$$

with

$$B = \begin{cases} \eta_{x(z-y)} & \text{when } \text{sign}\left(\frac{\omega_x}{\omega_y}\right) = -1 \\ \frac{1}{\eta_{x(y-z)}} & \text{when } \left(\frac{\omega_y}{\omega_x}\right) < 1 \text{ and } \text{sign}\left(\frac{\omega_x}{\omega_y}\right) = +1 \end{cases} \tag{21}$$

The proposed formula is more general than the one proposed by Radzimovsky (see Equation (25) of [26]).

The expressions of $\eta_{x(y-z)}$ and $\eta_{y(x-z)}$, for all possible combinations of x, y and z, are listed in Table 1 of reference [1] (see also Appendix A).

5. Numerical Example

We consider the case of the gear unit, shown in Figure 5, with x and y driving links and the z-driven link. In particular, $\omega_x = 8000$ rpm and $\omega_y = 7966.6$ rpm.

For a planet gear ratio $N_{j,i} = -1.5$,

$$\tau_1 = -\frac{1}{1 - N_{j,i}} = -0.4$$

and

$$\tau_2 = -\frac{N_{j,i}}{N_{j,i} - 1} = -0.6$$

Consequently, from Willis' equation, follows $\omega_z = 7980$ rpm.

Since $0 < \frac{\omega_x}{\omega_y} < 1$, the Formula (13) is applied. Considering the current value of $N_{j,i}$, and assuming

$$\eta_{z(x-y)} = 0.900$$

the entries 3a and 5a of Table A1 of the Appendix A give, respectively, the following:

$$\eta_{y(x-z)} = \frac{N_{j,i}\eta_{z(x-y)} - 1}{N_{j,i} - 1} = 0.940$$

$$\eta_{x(y-z)} = \frac{N_{j,i} - \eta_{z(x-y)}}{N_{j,i} - 1} = 0.960$$

For the prescribed numerical data, the Formula (13) yields the following:

$$\eta_{(x,y-z)} = \left| \frac{\omega_z \omega_{yx} \eta_{y(x-z)}}{\omega_x \omega_{zy} - \frac{\eta_{y(x-z)}}{\eta_{x(y-z)}} \omega_y \omega_{zx}} \right| = 0.998$$

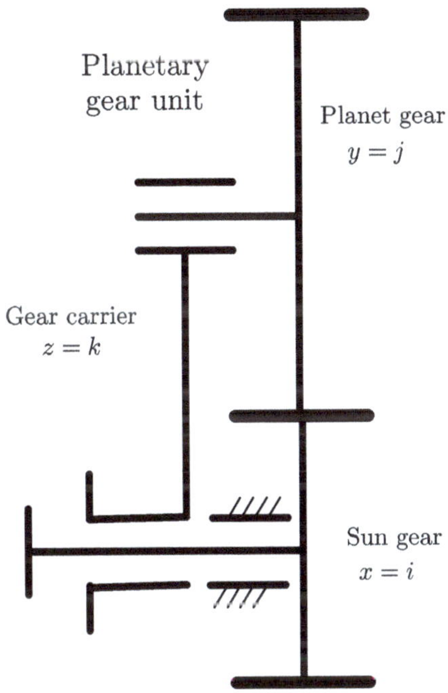

Figure 5. Kinematic structure of the PGU analyzed.

6. Conclusions

In this paper, the first novel result concerns the algebraic signs analysis of a virtual power-flow in a two dof PGU. It was shown that in a PGU, the virtual or potential powers, P_x^y and P_y^x, cannot have always the same algebraic sign. On the basis of this observation, new formulas for the computation of mechanical efficiency in a two dof PGU were deduced. The proposed formulas maintain their range of validity, also for cases not covered by the traditional Radzimovsky's formulas. Therefore, our approach offers a more general and compact alternative. The results herein can be extended also to other types of differential devices.

It must be observed that the current approach is limited to meshing losses only under a stationary working mode. Future work should embody in the formulas also the bearing losses. Moreover, the present treatment is based on stationary conditions; however, often, the efficiency under transient conditions is required. In this case, the time variation of kinetic energy due to inertia forces needs to be included in the power balance.

Author Contributions: Conceptualization, E.L.E. and E.P.; methodology, E.L.E., M.C. and E.P.; validation, E.L.E., M.C. and E.P.; writing—original draft preparation, E.L.E., M.C. and E.P.; writing—review and editing, E.L.E., M.C. and E.P. All authors have read and agreed to the published version of the manuscript.

Funding: This research received no external funding.

Institutional Review Board Statement: Not applicable.

Informed Consent Statement: Not applicable.

Data Availability Statement: Not applicable.

Conflicts of Interest: The authors declare no conflict of interest.

Nomenclature

dof	degree-of-freedom
$P_h = T_h \omega_h$	Power of link ($h = x, y, z$), greater/less than zero when h is a driving/driven link in the absolute motion;
$P_x^y = T_x(\omega_x - \omega_y)$	Potential or virtual power of link x when link y is considered fixed;
PGU	Planetary gear unit;
$N_{j,i} = \pm \frac{\text{No. teeth wheel } j}{\text{No. teeth wheel } i}$	Planet gear ratio (+: internally meshing gears, −: externally meshing gears)
T_x	Torque on link x
ω_x	Absolute angular velocity of link x
$\omega_{xy} = \omega_x - \omega_y$	Relative angular velocity of link x with respect to link y
$\eta_{x(y-z)}$	Mechanical efficiency of the one dof PGU when link x is fixed and links y and z are driving and driven links, respectively.
$\eta_{(x,y-z)}$	Mechanical efficiency of the two dof PGU when operating with links x and y as driving links and z as driven.
$\eta_{(z-x,y)}$	Mechanical efficiency of the two dof PGU when operating with links x and y as driven links and z as driving link.

Appendix A

In Table A1 are reported the efficiencies η of all possible kinematic inversions of a PGU [1].

The nomenclature of the table variables is as follows:

- i, j and k: subscripts denoting the gears and gear carrier, respectively;
- $R = N_{j,i}$: planet gear ratio;

Therefore, assuming $x = i$ (gear driving), $y = k$ (gear carrier driven), $z = j$ (fixed), $N_{j,i} = R > 1$, the mechanical efficiency of this gear arrangement is given by the entry 3a:

$$\eta_{z(x-y)} = \frac{R\eta_{k(i-j)} - 1}{R - 1}$$

The efficiency of the ordinary gear train can be accurately computed in different ways. However, for a crude approximation, the Tuplin formula can be used:

$$\eta_{k(i-j)} \approx \eta_{k(j-i)} \approx 1 - \left| \frac{1}{5} \left(\frac{1}{z_i} \pm \frac{1}{z_j} \right) \right| \tag{A1}$$

where the + and − sign applies for external or internal meshing gears, respectively.

Table A1. Efficiencies of single epicyclic spur-gear trains. $R = N_{j,i}$: planet gear ratio.

Case		Driving	Driven	Fixed	η
1		i	j	k	$\eta_{k(i-j)}$
2		j	i	k	$\eta_{k(j-i)}$
3a	$(R < 0)$ $(R > 1)$	i	k	j	$\frac{R\eta_{k(i-j)} - 1}{R-1}$
3b	$(0 < R < 1)$	i	k	j	$\frac{R - \eta_{k(j-i)}}{\eta_{k(j-i)}(R-1)}$
4a	$(R < 0)$ $(R > 1)$	k	i	j	$\frac{(R-1)\eta_{k(j-i)}}{R - \eta_{k(j-i)}}$
4b	$(0 < R < 1)$	k	i	j	$\frac{R-1}{R\eta_{k(i-j)} - 1}$
5a	$(R < 1)$	j	k	i	$\frac{R - \eta_{k(i-j)}}{R-1}$
5b	$(R > 1)$	j	k	i	$\frac{1 - R\eta_{k(i-j)}}{\eta_{k(i-j)}(1-R)}$
6a	$(R < 1)$	k	j	i	$\frac{(R-1)\eta_{k(i-j)}}{R\eta_{k(i-j)} - 1}$
6b	$(R > 1)$	k	j	i	$\frac{R-1}{R - \eta_{k(j-i)}}$

References

1. Pennestrì, E.; Mariti, L.; Valentini, P.P.; Mucino, V. Efficiency evaluation of gearboxes for parallel hybrid vehicles: Theory and applications. *Mech. Mach. Theory* **2012**, *49*, 157–176. [CrossRef]
2. Zhang, X.; Mi, C. *Vehicle Power Management—Modeling, Control and Optimization*; Springer: Berlin/Heidelberg, Germany, 2011.
3. Chen, C. Power flow analysis of compound epicyclic gear transmission: Simpson gear train. *ASME J. Mech. Des.* **2011**, *133*, 5. [CrossRef]
4. Chen, C. Power flow and efficiency analysis of epicyclic gear transmission with split power. *Mech. Mach. Theory* **2013**, *59*, 96–106. [CrossRef]
5. Esmail, E.L. Meshing efficiency analysis of two degree-of-freedom epicyclic gear trains. *ASME J. Mech. Des.* **2013**, *138*, 083301. [CrossRef]
6. Jianying, L.; Qingchun, H. Power Analysis and Efficiency Calculation of the Complex and Closed Planetary Gears Transmission. *Mech. Mach. Theory* **2016**, *100*, 423–433. [CrossRef]
7. Mohammadpour, M.; Theodossiades, S.; Rahnejat, H. Dynamics and efficiency of planetary gear sets for hybrid powertrains. *Proc. IMechE Part C J. Mech. Sci.* **2016**, *230*, 1359–1368. [CrossRef]
8. Coaccioli, P.; Pennestrì. Graph-Based Algorithm for the Evaluation of the Mechanical Efficiency of Epicyclic Gear Drive in Hybrid Scooters. In *Graph-Based Modelling in Engineering. Mechanisms and Machine Science*; Zawiślak, S., Rysiński, J., Eds.; Springer International Publishing: Berlin/Heidelberg, Germany, 2017; Volume 42, pp. 97–105.
9. Yang, F.; Feng, J.; Zhang, H. Power flow and efficiency analysis of multi-flow planetary gear trains. *Mech. Mach. Theory* **2015**, *92*, 86–99. [CrossRef]
10. Essam, E.L.; Juber, A.H.; Hussen, H.A. Power flow and efficiency analysis of multi-path transmission with planetary gear train. *Int. J. Mech. Eng. Technol.* **2018**, *9*, 412–423.
11. Hussen, H.A.; Esmail, E.L.; Hussen, R.A. Power Flow Simulation for Two-Degree-of-Freedom Planetary Gear Transmissions with Experimental Validation. *Model. Simul. Eng.* **2020**, *2020*, 14. [CrossRef]
12. De Carlo, M.; Mantriota, G. Electric vehicles with two motors combined via planetary gear train. *Mech. Mach. Theory* **2020**, *148*, 103789. [CrossRef]
13. Ding, H.; Cai, C.; Chen, Z.; Ke, T.; Mao, B. Configuration Synthesis and Performance Analysis of 9-Speed Automatic Transmissions. *Chin. J. Mech. Eng.* **2020**, *33*, 21. [CrossRef]
14. Mantriota, G.; Reina, G. Efficient Power Split Powertrain for Full Electric Vehicles. In *Mechanisms and Machine Science*; Springer International Publishing: Berlin/Heidelberg, Germany, 2020; pp. 560–567. [CrossRef]
15. Pennestrì, E.; Valentini, P.P. A review of formulas for the mechanical efficiency analysis of two degrees-of-freedom epicyclic gear trains. *ASME J. Mech. Des.* **2003**, *125*, 602–608. [CrossRef]
16. Macmillan, R.H. Epicyclic gear efficiencies. *Engineer* **1949**, *23*, 727–728.
17. Chen, C.; Angeles, J. Virtual-power flow and mechanical gear-mesh power losses of epicyclic gear trains. *J. Mech. Des. Trans. ASME* **2007**, *129*, 107–113. [CrossRef]
18. Rotella, D.; Cammalleri, M. Power losses in power-split CVTs: A fast black-box approximate method. *Mech. Mach. Theory* **2018**, *128*, 528–543. [CrossRef]

19. Esmail, E.L. Influence of the Operating Conditions of Two-Degree-of-Freedom Planetary Gear Trains on Tooth Friction Losses. *ASME J. Mech. Des.* **2018**, *140*, 054501. [CrossRef]
20. Malashchenko, V.; Strilets, O.; Strilets, V.; Kłlysz, S. Investigation of the energy effectiveness of multistage differential gears when the speed is changed by the carrier. *Diagnostyka* **2019**, *20*, 57–64. [CrossRef]
21. Cammalleri, M.; Castellano, A. Analysis of hybrid vehicle transmissions with any number of modes and planetary gearing: Kinematics, power flows, mechanical power losses. *Mech. Mach. Theory* **2021**, *162*, 104350. [CrossRef]
22. Sun, Z.; Gao, B.; Sanada, K. Power Loss Evaluation of Automated Manual Transmission with Gearshift Assistant Mechanism. *Int. J. Automot. Technol.* **2021**, *22*, 441–454. [CrossRef]
23. Chen, C.; Liang, T. Theoretic study of efficiency of two-DOFs of epicyclic gear transmission via virtual power. *ASME J. Mech. Des.* **2011**, *133*, 031007-1–031007-7. [CrossRef]
24. Chen, C. Power analysis of epicyclic transmissions based on constraints. *ASME J. Mech. Robot.* **2012**, *4*, 041004-1–041004-11. [CrossRef]
25. Radzimovsky, E.I. A Simplified Approach for Determing Power Losses and Efficiency of Planetary Gear Drives. *Mach. Des.* **1956**, *28*, 101–110.
26. Radzimovsky, E.I. How to find efficiency, speed and power losses in planetary gear drives. *Mach. Des.* **1959**, *31*, 144–153.
27. Pennestrì, E.; Freudenstein, F. The Mechanical Efficiency of Epicyclic Gear Trains. *ASME J. Mech. Des.* **1993**, *115*, 645–651. [CrossRef]
28. Esmail, E.L.; Pennestrì, E.; Hussein Juber, A. Power Losses in Two-Degrees-of-Freedom Planetary Gear Trains: A Critical Analysis of Radzimovsky's Formulas. *Mech. Mach. Theory* **2018**, *128*, 191–204. [CrossRef]
29. Pennestrì, E.; Freudenstein, F. A Systematic Approach to Power-Flow and Static Force Analysis in Epicyclic Spur-Gear Trains. *ASME J. Mech. Des.* **1993**, *115*, 639–644. [CrossRef]
30. Rabindran, D.; Tesar, D. Parametric design and power-flow analysis of parallel force/velocity actuators. *ASME J. Mech. Robot.* **2009**, *1*, 1–10. [CrossRef]

applied sciences

MDPI

Article

Performance Evaluation of a Compound Power-Split CVT for Hybrid Powertrains

Giacomo Mantriota *, Giulio Reina and Angelo Ugenti

Dipartimento di Meccanica, Matematica e Management, Politecnico di Bari, Via Orabona 4, 70125 Bari, Italy; giulio.reina@poliba.it (G.R.); angelo.ugenti@poliba.it (A.U.)
* Correspondence: giacomo.mantriota@poliba.it

Abstract: The Power-Split Continuously Variable Transmission is one of the most promising architectures for Hybrid Electric Vehicles. These systems have been introduced to improve vehicle global efficiency since they can maximize the efficiency in varying operating conditions. During the design stage, the availability of modeling tools would play a key role in achieving optimal design and control of these architectures. In this work, a compound power split device that combines an electric Continuously Variable Transmission with two planetary gear trains is analyzed. A comprehensive model is derived that allows the different power flow configurations to be evaluated given the properties of the single subcomponents of the system. The efficiency of the powertrain can be derived as well, and a numerical example is provided. The architecture studied has an efficiency that can be higher than that obtained using one single eCVT for most of the global transmission ratio range, showing that this solution could be suitable as a part of a more complex compound transmission that engages in a specific speed range.

Keywords: electric continuously variable transmission; power-split hybrid transmission; compound split transmission

Citation: Mantriota, G.; Reina, G.; Ugenti, A. Performance Evaluation of a Compound Power-Split CVT for Hybrid Powertrains. *Appl. Sci.* **2021**, *11*, 8749. https://doi.org/10.3390/app11188749

Academic Editor: Marco Cammalleri

Received: 23 August 2021
Accepted: 18 September 2021
Published: 20 September 2021

1. Introduction

Electrification of the powertrain is a big challenge for the current automotive industry [1]. One of the criticalities of this research area is the improvement of performance, fuel economy and driveability of Hybrid Electric Vehicles (HEV) [2]. Amongst the current hybrid powertrain architectures, the Power-Split Continuously Variable Transmission (PS-CVT) is the most promising technology, since HEVs often present multiple power sources that can be well managed by transmissions with multiple inputs [3].

Continuously Variable Transmissions (CVTs) were studied as early as late 19th century, when it was recognized that varying the drive ratio during vehicle operation would improve the performance [4]. Of the early designs, there have only been a few that have received significant attention from automotive designers. These CVTs can be classified into five categories, depending on their working principle: friction, traction, hydrostatic, electric, and variable geometry [5]. Conventional CVT units presents some limitations such as limited torque capacity and ratio coverage when compared to conventional units [6]. It has been proven that it is possible to overcome these limitations by combining the CVT with one or more epicyclic gears to create a power-split transmission [7].

An example of an early power-split design is the Perbury transmission proposed in 1940 and based on a toroidal variator (traction type CVT) and a planetary gear-set [8]. Interest in power-spit technology for automotive applications began in the 1980s. Products of this renewed interest are the Toyota Hybrid System and the Allison Hybrid Synergy by General Motors [9]. Since then, researchers have been attracted by this architecture [10]. Volpe et al. [11] presented an optimization procedure for designing the input and output coupled PS-CVT with higher efficiency. Kim et al. [12] proposed a two-mode PS-CVT with a mode control strategy to minimize fuel consumption. Zhang et al. [13] explored potential compound power-split configurations with two planetary gear sets to understand how to design the best compound power-split configuration.

The PS-CVT architectures can be classified as single-mode (i.e., input split or output split) or multi-mode, in which the power split type is selected engaging and disengaging clutches included in the transmission layout. In general, single-mode PS-CVTs show low efficiency at high speed when compared to more complex solutions that include at least two epicyclical gear trains and one or more locking systems [14–16]. These solutions are also called compound type power split and an example is given by Global Hybrid Cooperation, a joint project by GM, Daimler, and BMW [17].

Our research group has recently studied the power flows and efficiency of the PS-CVT [18–20]. In these previous works, the analysis of a "four-port-mechanical-power split device" was proposed. This device consists of an electric Continuously Variable Transmission (eCVT) and a power split device (Figure 1). The former is obtained connecting two electric motors; the latter is composed by two planetary gear sets named PG1 and PG2. Figure 1 shows the four possible set-ups. While types A, B and D have been discussed in [18–20], in this paper, we focus on type C. The power flow and the efficiency of this "four-port-mechanical-power split device" are defined and its performance assessed with a numerical example.

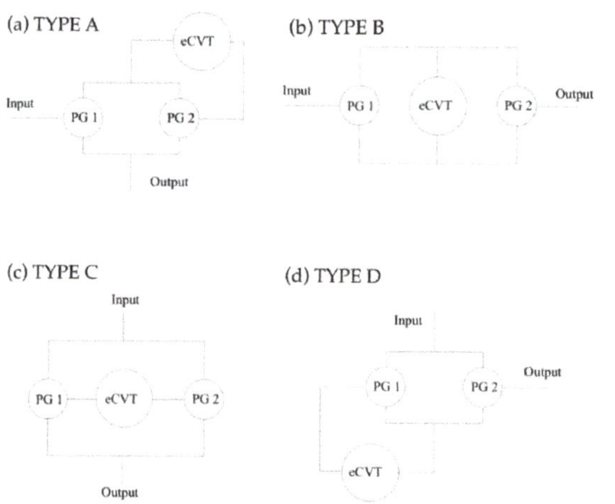

Figure 1. Compound eCVT based on the four-port-mechanical power-split device: (**a**) type A arrangement; (**b**) type B arrangement; (**c**) type C arrangement; (**d**) type D arrangement.

The practical value of this paper is to provide the designer of such architecture with a tool to estimate advantages and disadvantages of implementing this topology within a powertrain, e.g., the type of power flow that will occur, the power that will flow in the eCVT and the overall efficiency, knowing just the design parameters, that are the geometry of the planetary gear trains, the ratio spread of the eCVT and the required overall speed ratio spread.

Section 2 shows the formulation of the kinematic model of the transmission. The types of power flows and the specific conditions at which they occur are presented in Section 3 together with the derivation of formulas for the efficiency of the whole system. Section 4 provides a numerical example in which the performance of the transmission is estimated for specific working conditions. Finally, the conclusions of the work are drawn in Section 5.

2. Kinematic Model of a Compound Power-Split eCVT

The compound power-split eCVT analyzed in this paper is shown in Figure 2 and it is one of the possible arrangements of the "four-port-mechanical-power split device". The shafts are numbered from 1 to 6. In this set-up, the shafts that drive the central gears of the

PGs (shafts 1, 2, 3, and 4) are connected to the input and output shafts, while the planet carriers (shafts 5 and 6) are connected to the eCVT. At this stage, it is not specified which one of the two central gears (i.e., sun and ring) of PG1 is connected to shaft 1 or shaft 2, and the same applies to PG2. It is a choice of the designer that depends on the value of the transmission ratio required by the specific application.

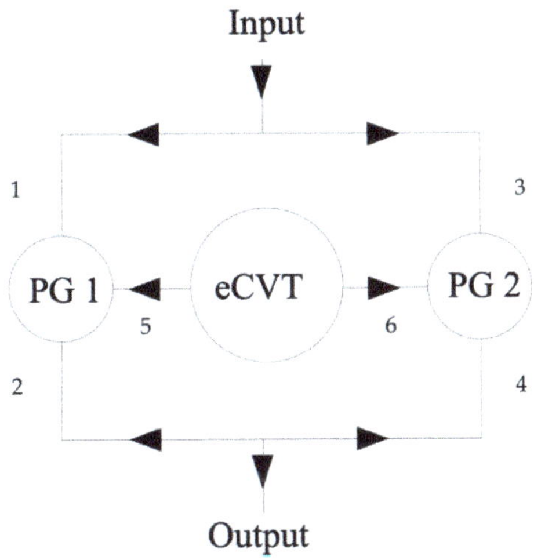

Figure 2. Shafts indexing with positive direction of power on all shafts.

The transmission ratios for the PGs and the eCVT are given by (using Willis equation for PG1 and PG2 [21]):

$$\tau_{W1} = \frac{\omega_2 - \omega_5}{\omega_1 - \omega_5}; \qquad \tau_{W2} = \frac{\omega_4 - \omega_6}{\omega_3 - \omega_6}; \qquad \tau_{eCVT} = \frac{\omega_6}{\omega_5} \tag{1}$$

with:

$$\omega_1 = \omega_3 = \omega_i; \qquad \omega_2 = \omega_4 = \omega_o \tag{2}$$

where ω_j represents the angular speed of shaft j, ω_i is the input speed, and ω_o is the output speed. It is important to specify that the planetary gear ratios τ_{W1} and τ_{W2} are design parameters and therefore constants defined by the geometry of the gears. On the other hand, the global and eCVT speed ratios are kinematic variables, and the design choice regards their spread rather than their instantaneous value that depends on the input and output speeds.

The following speed ratios can be derived from Equations (1) and (2):

$$\frac{\omega_5}{\omega_i} = \frac{\tau_{W2} - \tau_{W1}}{\tau_{eCVT}(\tau_{W2} - 1) - (\tau_{W1} - 1)}; \qquad \frac{\omega_6}{\omega_i} = \frac{\tau_{eCVT}(\tau_{W2} - \tau_{W1})}{\tau_{eCVT}(\tau_{W2} - 1) - (\tau_{W1} - 1)} \tag{3}$$

$$\frac{\omega_5}{\omega_o} = \frac{\tau_{W2} - \tau_{W1}}{\tau_{eCVT}\tau_{W1}(\tau_{W2} - 1) - \tau_{W2}(\tau_{W1} - 1)}; \qquad \frac{\omega_6}{\omega_o} = \frac{\tau_{eCVT}(\tau_{W2} - \tau_{W1})}{\tau_{eCVT}\tau_{W1}(\tau_{W2} - 1) - \tau_{W2}(\tau_{W1} - 1)} \tag{4}$$

Rearranging and combining Equations (1) and (2), the global speed ratio can be derived as a function of the transmission ratios of PGs and eCVT:

$$\tau_{gl} = \frac{\omega_O}{\omega_i} = \frac{\omega_2}{\omega_1} = \frac{\tau_{W1}\tau_{eCVT}(\tau_{W2} - 1) - \tau_{W2}(\tau_{W1} - 1)}{\tau_{eCVT}(\tau_{W2} - 1) - (\tau_{W1} - 1)} \tag{5}$$

The analytical expression of the partial derivative of the global transmission ratio with respect to the speed ratio of the eCVT is derived from Equation (5):

$$\frac{\partial \tau_{gl}}{\partial \tau_{eCVT}} = \frac{(\tau_{W2}-1)(\tau_{W1}-\tau_{W2})(1-\tau_{W1})}{(\tau_{eCVT}(\tau_{W2}-1)-(\tau_{W1}-1))^2} \tag{6}$$

The sign of Equation (6) varies according to the numerator. Therefore, the function $\tau_{gl}(\tau_{eCVT})$ is monotonic for a given set of planetary gear trains. The type of proportionality (namely direct or indirect) between τ_{gl} and τ_{eCVT} is a design choice. Hereafter, a direct proportionality is assumed, meaning that the following condition must be verified:

$$(\tau_{W2}-1)(\tau_{W1}-\tau_{W2})(1-\tau_{W1}) > 0 \tag{7}$$

The minimum and maximum values of the global speed ratio (namely τ_{gl_m} and τ_{gl_M}, respectively) are given by the specific application for which the transmission is designed. These values are related to the limits of the eCVT speed ratio (namely τ_{eCVTm} and τ_{eCVTM}) via Equation (5):

$$\tau_{gl_m} = \frac{\tau_{W1}\tau_{eCVTm}\,(\tau_{W2}-1)-\tau_{W2}(\tau_{W1}-1)}{\tau_{eCVTm}(\tau_{W2}-1)-(\tau_{W1}-1)} \tag{8}$$

$$\tau_{gl_M} = \frac{\tau_{W1}\tau_{eCVTM}\,(\tau_{W2}-1)-\tau_{W2}(\tau_{W1}-1)}{\tau_{eCVTM}(\tau_{W2}-1)-(\tau_{W1}-1)} \tag{9}$$

The maximum and minimum eCVT speed ratios are design inputs as well. For this reason, Equations (8) and (9) are useful in the early design stage to determine the characteristics of the planetary gear trains. Indeed, these expressions can be rearranged solving with respect to τ_{W1} and τ_{W2}:

$$\tau_{W1} = \frac{\tau_{Glm}\tau_{eCVTm}\left(1-\tau_{gl_M}\right)-\tau_{gl_M}\tau_{eCVTM}\left(1-\tau_{gl_m}\right)+\left(\tau_{gl_M}-\tau_{gl_m}\right)}{\tau_{eCVTm}\left(1-\tau_{gl_M}\right)+\tau_{gl_m}(\tau_{eCVTM}-1)+\left(\tau_{gl_M}-\tau_{eCVTM}\right)} \tag{10}$$

$$\tau_{W2} = \frac{\tau_{gl_m}\tau_{eCVTm}\left(\tau_{gl_M}-\tau_{eCVTM}\right)-\tau_{gl_M}\tau_{eCVTM}\left(\tau_{gl_m}-\tau_{eCVTm}\right)+\tau_{eCVTM}\tau_{gl_m}-\tau_{gl_M}\tau_{eCVTm}}{\tau_{eCVTm}\tau_{gl_m}(1-\tau_{eCVTM})+\tau_{gl_M}\tau_{eCVTM}(\tau_{eCVTm}-1)+(\tau_{eCVTM}-\tau_{eCVTm})} \tag{11}$$

Similar expressions can be easily derived for the case in which indirect proportionality between τ_{gl} and τ_{eCVT} is adopted, by imposing:

$$\tau_{gl_M} = \frac{\tau_{W1}\tau_{eCVTm}\,(\tau_{W2}-1)-\tau_{W2}(\tau_{W1}-1)}{\tau_{eCVTm}(\tau_{W2}-1)-(\tau_{W1}-1)} \tag{12}$$

$$\tau_{gl_m} = \frac{\tau_{W1}\tau_{eCVTM}\,(\tau_{W2}-1)-\tau_{W2}(\tau_{W1}-1)}{\tau_{eCVTM}(\tau_{W2}-1)-(\tau_{W1}-1)} \tag{13}$$

3. Power Flows

The compound power split devices are subject to several types of internal power flows because of the presence of two planetary gear sets. The objective of this section is to derive the specific conditions at which each power flow occurs in the type C "four-port-mechanical-power split device". The conditions for each power flow in terms of planetary gears speed ratios are studied in two cases, depending on the sign of the global transmission ratio. The schematics of the twelve possible types of power flows are shown in Figure 3, while the conditions for each power flow type in terms of power ratio signs are described in Table 1, where P_j is the power across shaft j. In this paper, P_j is assumed positive when entering the planetary gears. The positive direction of power for all shafts is shown in Figure 2.

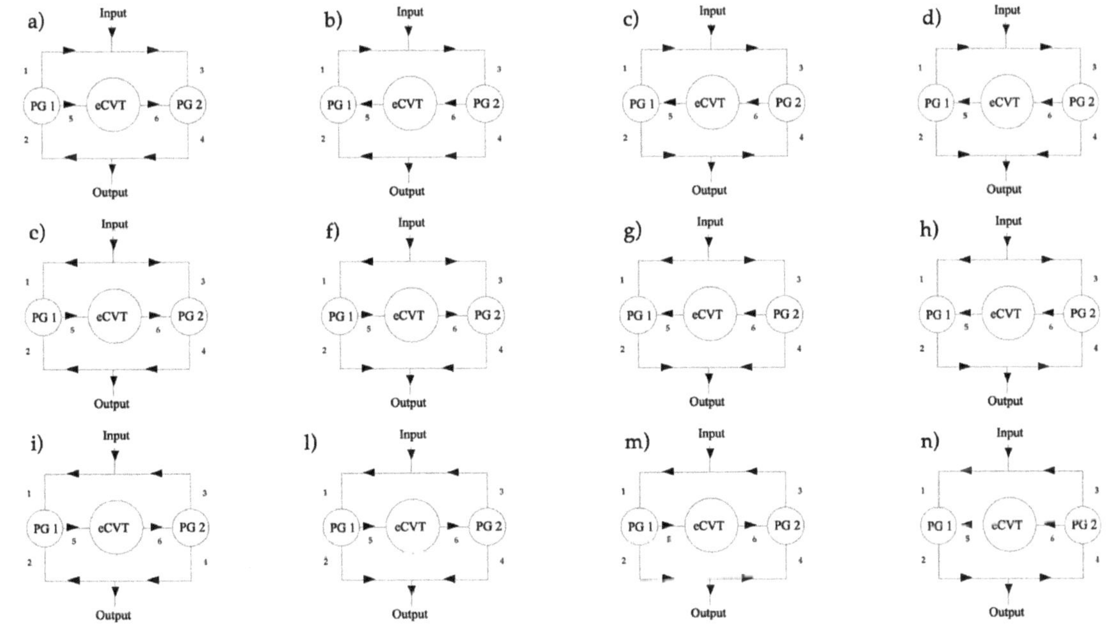

Figure 3. Power flow types within a type C compound transmission: twelve possible flows with different power circulations.

Table 1. Power ratio signs required to obtain each power flow type.

Power Flow Type (Figure 3)	$\frac{P_1}{P_2}$	$\frac{P_3}{P_4}$	$\frac{P_1}{P_5}$	$\frac{P_3}{P_6}$
(a)	< 0	< 0	> 0	> 0
(b)	< 0	< 0	< 0	< 0
(c)	> 0	> 0	< 0	< 0
(d)	> 0	< 0	< 0	< 0
(e)	> 0	< 0	< 0	> 0
(f)	< 0	< 0	< 0	> 0
(g)	< 0	< 0	> 0	< 0
(h)	< 0	> 0	> 0	< 0
(i)	> 0	> 0	< 0	< 0
(l)	< 0	> 0	< 0	< 0
(m)	< 0	< 0	< 0	< 0
(n)	< 0	< 0	> 0	> 0

To obtain analytical expressions for the power ratios across the various shafts, the torque ratios must be computed first. Assuming no losses in the planetary gear trains, the following equations can be derived:

$$\frac{T_1}{T_2} = -\tau_{W1}; \quad \frac{T_3}{T_4} = -\tau_{W2}; \quad \frac{T_1}{T_5} = \frac{\tau_{W1}}{1-\tau_{W1}}; \quad \frac{T_3}{T_6} = \frac{\tau_{W2}}{1-\tau_{W2}} \tag{14}$$

From Equations (3), (4), and (14), the following expressions for the power ratios can be obtained:

$$\frac{P_1}{P_2} = \frac{T_1\omega_i}{T_2\omega_o} = -\frac{\tau_{W1}}{\tau_{gl}}; \quad \frac{P_3}{P_4} = \frac{T_3\omega_i}{T_4\omega_o} = -\frac{\tau_{W2}}{\tau_{gl}}; \quad \frac{P_1}{P_5} = \frac{T_1\omega_i}{T_5\omega_5} = \frac{\tau_{W1}}{\tau_{gl}-\tau_{W1}}; \quad \frac{P_3}{P_6} = \frac{T_3\omega_i}{T_6\omega_6} = \frac{\tau_{W2}}{\tau_{gl}-\tau_{W2}} \tag{15}$$

The sign of the power ratios is not enough to uniquely determine the type of power flow. For example, types a and n present the same combination of power ratio signs, and therefore a further condition is required to distinguish them. The difference between power flows (a) and (n) is in the direction of the power across shafts 1 and 3: power flow (a) occurs when $|T_3| > |T_1|$. From Equation (14), it can be obtained:

$$\frac{T_3}{T_6}\frac{T_5}{T_1} = \frac{\tau_{W2}(1-\tau_{W1})}{(1-\tau_{W2})\tau_{W1}} \tag{16}$$

From Equations (1) and (16) it follows:

$$\frac{T_3}{T_1} = \frac{\tau_{W2}(1-\tau_{W1})}{\tau_{eCVT}\tau_{W1}(1-\tau_{W2})} \tag{17}$$

From Equation (17) it can be concluded that power flow type (a) occurs when:

$$|\tau_{eCVT}| < \left|\frac{\tau_{W2}(1-\tau_{W1})}{\tau_{W1}(1-\tau_{W2})}\right| \tag{18}$$

The same consideration applies to power flows (b)–(m) and (c)–(i).

In the next two sections, the sign of the power ratios of Equation (15) and the conditions for each power flow are studied in two cases:

- Case 1: $\tau_{gl} > 0$;
- Case 2: $\tau_{gl} < 0$.

3.1. Case 1: $\tau_{gl} > 0$

In the case the global speed ratio is positive, Table 2 shows the ranges of τ_{W1} and τ_{W2} for each possible power ratio sign as derived from Equation (15). Combining the information in Tables 1 and 2, the conditions for each power flow are obtained and shown in Table 3 where the additional condition refers to Equation (18). These results can be used to uniquely determine the type of power flow that occurs in a specific application of the "four-port-mechanical-power split device". For example, considering a device designed with $\tau_{W1} < 0$ working in a condition in which $\tau_{W2} > \tau_{gl}$, the resulting power flow will be of type (d).

Table 2. Case 1—Ranges of planetary gears speed ratios for each power ratio sign.

Power Ratio	>0	<0
P_1/P_2	$\tau_{W1} < 0$	$\tau_{W1} > 0$
P_3/P_4	$\tau_{W2} < 0$	$\tau_{W2} > 0$
P_1/P_5	$0 < \tau_{W1} < \tau_{gl}$	$\tau_{W1} < 0$ or $\tau_{W1} > \tau_{gl}$
P_3/P_6	$0 < \tau_{W2} < \tau_{gl}$	$\tau_{W1} < 0$ or $\tau_{W2} > \tau_{gl}$

3.2. Case 2: $\tau_{gl} < 0$

Similar considerations can be repeated when the global speed ratio is negative. For this case, Table 4 shows the ranges of τ_{W1} and τ_{W2} for each possible power ratio sign as derived from Equation (15). Combining the Tables 1 and 4, the conditions for each power flow are obtained and shown in Table 5 where the additional condition refers to Equation (18).

Table 3. Case 1—Conditions for each power flow.

Power Flow Type (Figure 3)	τ_{W1}	τ_{W2}	Additional Condition
(a)	$0 < \tau_{W1} < \tau_{gl}$	$0 < \tau_{W2} < \tau_{gl}$	$\lvert\tau_{eCVT}\rvert < \left\lvert\frac{\tau_{W2}(1-\tau_{W1})}{\tau_{W1}(1-\tau_{W2})}\right\rvert$
(b)	$\tau_{W1} > \tau_{gl}$	$\tau_{W2} > \tau_{gl}$	$\lvert\tau_{eCVT}\rvert < \left\lvert\frac{\tau_{W2}(1-\tau_{W1})}{\tau_{W1}(1-\tau_{W2})}\right\rvert$
(c)	$\tau_{W1} < 0$	$\tau_{W2} < 0$	$\lvert\tau_{eCVT}\rvert < \left\lvert\frac{\tau_{W2}(1-\tau_{W1})}{\tau_{W1}(1-\tau_{W2})}\right\rvert$
(d)	$\tau_{W1} < 0$	$\tau_{W2} > \tau_{gl}$	
(e)	$\tau_{W1} < 0$	$0 < \tau_{W2} < \tau_{gl}$	
(f)	$\tau_{W1} > \tau_{gl}$	$0 < \tau_{W2} < \tau_{gl}$	
(g)	$0 < \tau_{W1} < \tau_{gl}$	$\tau_{W2} > \tau_{gl}$	
(h)	$0 < \tau_{W1} < \tau_{gl}$	$\tau_{W2} < 0$	
(i)	$\tau_{W1} < 0$	$\tau_{W2} < 0$	$\lvert\tau_{eCVT}\rvert > \left\lvert\frac{\tau_{W2}(1-\tau_{W1})}{\tau_{W1}(1-\tau_{W2})}\right\rvert$
(l)	$\tau_{W1} > \tau_{gl}$	$\tau_{W2} < 0$	
(m)	$\tau_{W1} > \tau_{gl}$	$\tau_{W2} > \tau_{gl}$	$\lvert\tau_{eCVT}\rvert > \left\lvert\frac{\tau_{W2}(1-\tau_{W1})}{\tau_{W1}(1-\tau_{W2})}\right\rvert$
(n)	$0 < \tau_{W1} < \tau_{gl}$	$0 < \tau_{W2} < \tau_{gl}$	$\lvert\tau_{eCVT}\rvert > \left\lvert\frac{\tau_{W2}(1-\tau_{W1})}{\tau_{W1}(1-\tau_{W2})}\right\rvert$

Table 4. Case 2—Ranges of planetary gears speed ratios for each power ratio sign.

Power Ratio	>0	<0
P_1/P_2	$\tau_{W1} > 0$	$\tau_{W1} < 0$
P_3/P_4	$\tau_{W2} > 0$	$\tau_{W2} < 0$
P_1/P_5	$\tau_{gl} < \tau_{W1} < 0$	$\tau_{W1} < \tau_{gl}$ or $\tau_{W1} > 0$
P_3/P_6	$\tau_{gl} < \tau_{W2} < 0$	$\tau_{W1} < \tau_{gl}$ or $\tau_{W1} > 0$

Table 5. Case 2—Conditions for each power flow.

Power Flow Type (Figure 3)	τ_{W1}	τ_{W2}	Additional Condition
(a)	$\tau_{gl} < \tau_{W1} < 0$	$\tau_{gl} < \tau_{W2} < 0$	$\lvert\tau_{eCVT}\rvert < \left\lvert\frac{\tau_{W2}(1-\tau_{W1})}{\tau_{W1}(1-\tau_{W2})}\right\rvert$
(b)	$\tau_{W1} < \tau_{gl}$	$\tau_{W2} < \tau_{gl}$	$\lvert\tau_{eCVT}\rvert < \left\lvert\frac{\tau_{W2}(1-\tau_{W1})}{\tau_{W1}(1-\tau_{W2})}\right\rvert$
(c)	$\tau_{W1} > 0$	$\tau_{W2} > 0$	$\lvert\tau_{eCVT}\rvert < \left\lvert\frac{\tau_{W2}(1-\tau_{W1})}{\tau_{W1}(1-\tau_{W2})}\right\rvert$
(d)	$\tau_{W1} > 0$	$\tau_{W2} < \tau_{gl}$	
(e)	$\tau_{W1} > 0$	$\tau_{gl} < \tau_{W2} < 0$	
(f)	$\tau_{W1} < \tau_{gl}$	$\tau_{gl} < \tau_{W2} < 0$	
(g)	$\tau_{gl} < \tau_{W1} < 0$	$\tau_{W2} < \tau_{gl}$	
(h)	$\tau_{gl} < \tau_{W1} < 0$	$\tau_{W2} > 0$	
(i)	$\tau_{W1} > 0$	$\tau_{W2} > 0$	$\lvert\tau_{eCVT}\rvert > \left\lvert\frac{\tau_{W2}(1-\tau_{W1})}{\tau_{W1}(1-\tau_{W2})}\right\rvert$
(l)	$\tau_{W1} < \tau_{gl}$	$\tau_{W2} > 0$	
(m)	$\tau_{W1} < \tau_{gl}$	$\tau_{W2} < \tau_{gl}$	$\lvert\tau_{eCVT}\rvert > \left\lvert\frac{\tau_{W2}(1-\tau_{W1})}{\tau_{W1}(1-\tau_{W2})}\right\rvert$
(n)	$\tau_{gl} < \tau_{W1} < 0$	$\tau_{gl} < \tau_{W2} < 0$	$\lvert\tau_{eCVT}\rvert > \left\lvert\frac{\tau_{W2}(1-\tau_{W1})}{\tau_{W1}(1-\tau_{W2})}\right\rvert$

3.3. *Efficiency*

It has been demonstrated by Cammalleri et al. [14–16] that, ignoring losses, the power through the eCVT divided by the power in input to the compound power-split device is given by the following formula, independently of the internal arrangement:

$$\left|\frac{P_{eCVT}}{P_i}\right|_{ideal} = \left|\frac{\partial \tau_{Gl}}{\partial \tau_{eCVT}}\frac{\tau_{eCVT}}{\tau_{Gl}}\right| = \frac{\tau_{eCVT}(\tau_{W2}-1)(\tau_{W1}-\tau_{W2})(1-\tau_{W1})}{(\tau_{eCVT}(\tau_{W2}-1)-(\tau_{W1}-1))(\tau_{W1}\tau_{eCVT}(\tau_{W2}-1)-\tau_{W2}(\tau_{W1}-1))} \tag{19}$$

In a real case, the power loss in the transmission is mainly due to the eCVT. The overall efficiency can be derived under the hypothesis that the power losses in the other components are negligible. This assumption is introduced because the efficiency of the eCVT is generally very low compared to the efficiency of the planetary gear trains. With this assumption, the global efficiency can be calculated as follows

$$\eta_{gl} = \frac{P_i + P_{Loss}}{P_i} \tag{20}$$

The conservation of energy applied to the node connecting the input to shafts 1 and 3 leads to:

$$P_i - P_1 - P_3 = 0 \tag{21}$$

If the power in the eCVT flows from left to right, that is when $P_5 < 0$ and $P_6 > 0$ (power flows (a), (e), (f), (i), (l) and (m)), the efficiency of the eCVT is given by:

$$\eta_{eCVT} = -\frac{P_6}{P_5} \tag{22}$$

where the signs of the powers refer to Figure 2. From the conservation of energy across the eCVT, the power loss in the eCVT is given by:

$$P_{Loss} = P_5 + P_6 = (1 - \eta_{eCVT})P_5 \tag{23}$$

Combining Equations (15), (20)–(23) the global efficiency can be derived as a function of the characteristic transmission ratios and of the efficiency of the eCVT:

$$\eta_{gl} = \tau_{gl} \frac{\tau_{gl} - \tau_{W2} - \eta_{eCVT}\left(\tau_{gl} - \tau_{W1}\right)}{\tau_{W1}\left(\tau_{gl} - \tau_{W2}\right) - \eta_{eCVT}\tau_{W2}\left(\tau_{gl} - \tau_{W1}\right)} \tag{24}$$

If the power in the eCVT flows from right to left ($P_5 > 0$ and $P_6 < 0$, power flows (b), (c), (d), (g), (h) and (n)), the efficiency of the eCVT is given by:

$$\eta_{eCVT} = -\frac{P_5}{P_6} \tag{25}$$

From the conservation of energy across the eCVT, the power loss in the eCVT is given by:

$$P_{Loss} = P_5 + P_6 = -\frac{(1 - \eta_{eCVT})}{\eta_{eCVT}} P_5 \tag{26}$$

Combining Equations (15), (22), (23), (25) and (26) the global efficiency is given by:

$$\eta_{gl} = \tau_{gl} \frac{\tau_{gl} - \tau_{W1} - \eta_{eCVT}\left(\tau_{gl} - \tau_{W2}\right)}{\tau_{W2}\left(\tau_{gl} - \tau_{W1}\right) - \eta_{eCVT}\tau_{W1}\left(\tau_{gl} - \tau_{W2}\right)} \tag{27}$$

In conclusion, the purpose of this approach is to develop a tool to study the advantages and disadvantages of the type C compound power-split device and to evaluate its performance and the power flow obtained within the system. In the design stage, knowing the input design parameters (namely the ranges of global and eCVT speed ratios), the parameters of the planetary gears are obtained with Equations (10) and (11) (assuming direct proportionality between global and eCVT speed ratios). Once the components of the transmission are designed, the Tables 3 and 5 show the type of power flow according to the kinematic parameters of the transmission. Equations (19), (24), and (27) can be used to estimate the power flowing through the eCVT and the efficiency of the whole system.

4. Numerical Example

In this section the performance of the type C "four-port-mechanical-power split device" (Figure 1c) is evaluated through a numerical example. This kind of architecture is often used as part of a more complex compound transmission and engaged in a specific speed range, as for example shown by Zhang et al. [13]. Assuming that we are designing a complex compound power split transmission that changes configuration depending on the speed range, we can decide to engage the topology presented in this paper when the global transmission ratio is between 0.7 and 1.5.

The input design parameters are therefore chosen as follows:

$$\tau_{gl_m} = 0.7, \qquad \tau_{gl_M} = 1.5, \qquad \tau_{eCVTm} = -1.3, \qquad \tau_{eCVTM} = 1.3 \tag{28}$$

Assuming inverse proportionality between global and eCVT speed ratios, the corresponding values of the planetary gear ratios are calculated with Equations (12) and (13):

$$\tau_{W1} = 1.722, \qquad \tau_{W2} = 1.357 \tag{29}$$

Figure 4 shows how the global transmission ratio varies as a function of the speed ratio of the eCVT (Equation (5)).

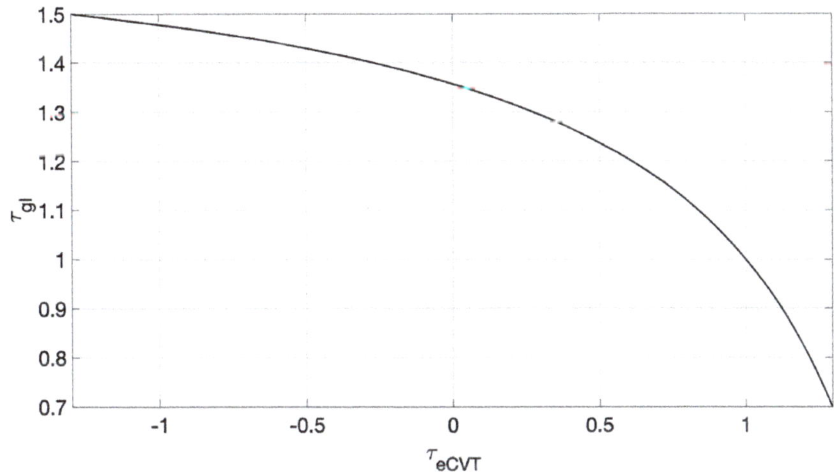

Figure 4. Global transmission ratio as a function of eCVT transmission ratio, given the input design parameters of Equations (28) and (29).

Although the input parameters obtained so far appear reasonable for conventional eCVT and planetary gear trains, it is important to assess the feasibility of the application by checking the speed of the two electric machines in the eCVT (namely, ω_5 and ω_6). Assuming a constant input speed of 4000 RPM, Figure 5 shows the speed of the electric machines (Equation (4)), as a function of the global transmission ratio. The highest speed is reached by the electric machine connected to shaft 6, which spins at a rate of 7360 RPM when the global transmission ratio is 0.7. Considering that conventional electric machines can work up to 9000 RPM [22], it is safe to accept these speed ranges.

The fraction of power through the eCVT in the ideal case $|P_{eCVT}/P_i|_{ideal}$ is plotted in Figure 6 as a function of the global transmission ratio, as obtained from Equation (19). The efficiency of the whole system is shown in Figure 7 as a function of the global transmission ratio, where the efficiency of the eCVT is assumed equal to 0.9.

In the plots of Figures 6 and 7, two different areas can be identified and each one corresponds to a different type of power flow. For values of the global speed ratio between 1.357 and 1.5 power flow (f) is established in the transmission (Figure 3f). In this case, the efficiency is calculated with Equation (24) because the power through the eCVT flows from left to right. For τ_{gl} between 0.7 and 1.357, power flow (b) occurs, and the efficiency is calculated with Equation (27) because the power through the eCVT has changed direction, flowing from right to left.

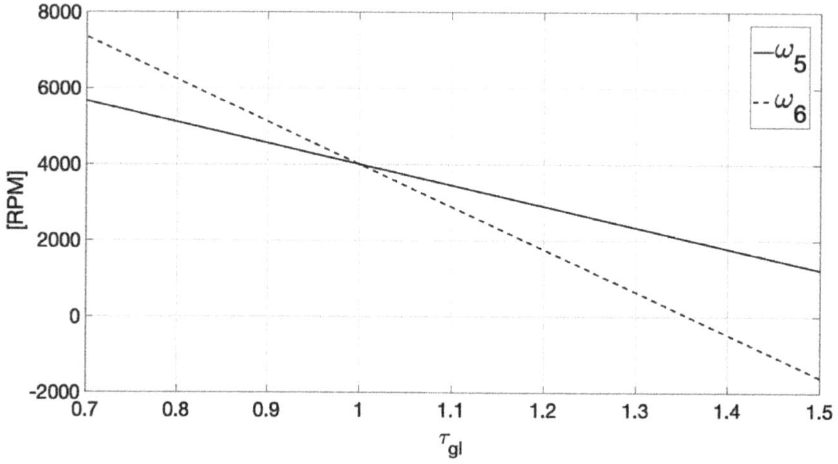

Figure 5. Speed of the electric machines as a function of the global transmission ratio.

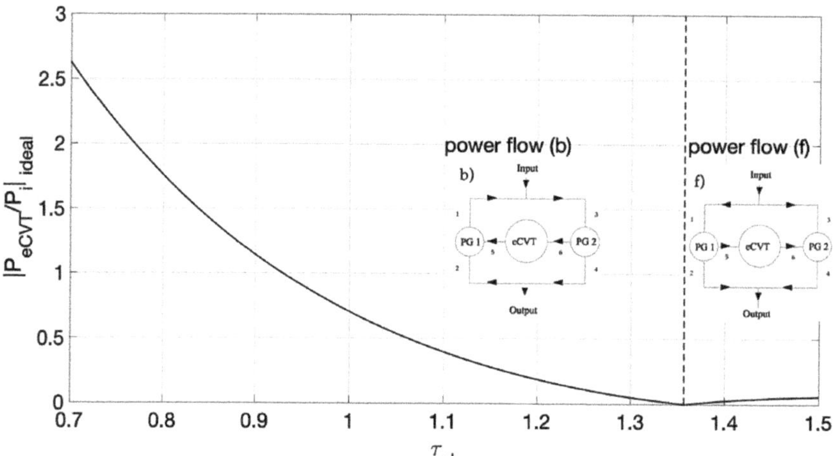

Figure 6. eCVT power fraction vs. global transmission ratio, given the input design parameters of Equations (28) and (29).

The use of the type C architecture for this application is justified by the improved efficiency of the whole system, which is much higher than the efficiency of the eCVT for most of the range of the global transmission ratio ($0.88 < \tau_{gl} < 1.5$). This is explained by the fact that in the same region, the power flowing through the eCVT, that is the only source of power losses, is lower than the input power ($|P_{eCVT}/P_i|_{ideal} < 1$). This condition is verified by definition for the case of power flow (f), because the input power splits twice before entering the eCVT, first between shafts 1 and 3, and then between shafts 5 and 2 Figure 3f). This means that power flow (f) is always advantageous. In the case of power flow (b), the power through the eCVT is low when the global transmission ratio is between 0.88 and 1.357, indicating that no recirculation occurs. However, when the global transmission ratio goes below 0.88 a significant fraction of the input power recirculates through the eCVT, making the overall efficiency drop.

The drop in efficiency for low values of the global transmission ratio suggests that this topology would be better suited as a part of a more complex compound transmission, in which a system of clutches and brakes allows the user to change the architecture, choosing one of the four configurations (Figure 1), connecting shafts 1, 2, 3, 4, 5, and 6 accordingly.

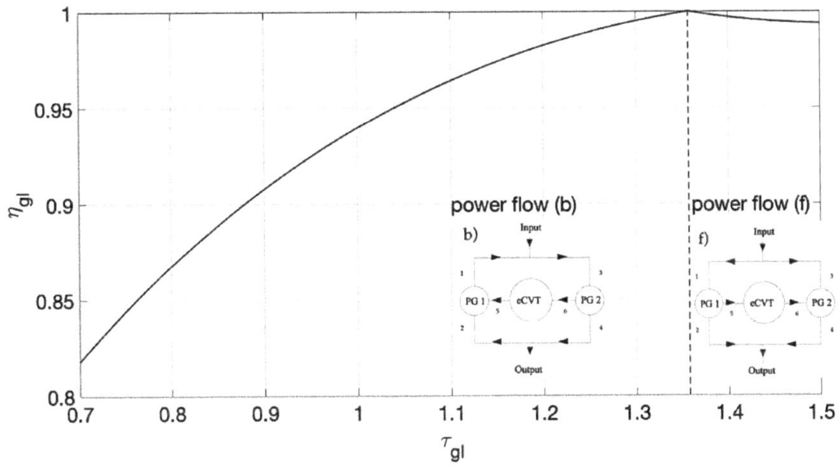

Figure 7. Global efficiency vs. global transmission ratio, given the input design parameters of Equations (28) and (29).

5. Conclusions

This work focused on the evaluation of a compound power split device. The analyzed transmission module is a type C four-port power split device. The twelve types of possible power flow configurations were studied and the conditions at which each power flow occurs have been derived. The main motivation behind this research was to study the selected transmission topology and its properties, which will allow designers and practitioners to estimate advantages and disadvantages of implementing this architecture within a complex compound powertrain. A numerical example was proposed in which a type C compound power split transmission is used. It was concluded that, although the efficiency of the system is higher than the efficiency of the eCVT for most of the global transmission ratio range, this topology would be better suited as a part of a more complex compound transmission to be engaged in a specific speed range.

This work completes the in depth study of the possible configurations of the four-port power split device. Knowing the bounds of the eCVT and global speed ratios, the models developed in these works establish an original approach by which is possible to evaluate:

- The type of power flow
- The amount of power that re-circulates through the e-CVT
- The overall efficiency of the system.

The main asset of the four-port power split device is that it is possible to design a system of clutches and brakes that allows the user to change the architecture, choosing one of the four configurations, connecting shafts 1, 2, 3, 4, 5, and 6 accordingly. The objective for future works is to develop an optimization tool that identifies the most efficient configuration for each working condition. Ideally, this tool will help designers to choose the best kinematic parameters (eCVT and planetary gears speed ratios) and obtain an optimum four-port power split device for each specific application.

Author Contributions: All authors certify that they have participated sufficiently in the work to take public responsibility for the content, including participation in the concept, design, analysis, writing, or revision of the manuscript. All authors have read and agreed to the published version of the manuscript.

Funding: This research received no external funding.

Institutional Review Board Statement: Not applicable.

Informed Consent Statement: Not applicable.

Data Availability Statement: Not applicable.

Acknowledgments: Politecnico di Bari is gratefully acknowledged for the useful contributions in the development analysis.

Conflicts of Interest: The authors declare no conflict of interest.

References

1. Muniamuthu, S.; Krishna Arjun, S.; Jalapathy, M.; Harikrishnan, S.; Vignesh, A. Review on Electric Vehicles. *Int. J. Mech. Prod. Eng. Res. Dev.* **2018**, *8*, 557–566. [CrossRef]
2. Huang, Y.; Surawski, N.; Organ, B.; Zhou, J.L.; Tang, O.H.; Chan, E.F. Fuel consumption and emissions performance under real driving: Comparison between hybrid and conventional vehicles. *Sci. Total. Environ.* **2018**, *659*, 275–282. [CrossRef] [PubMed]
3. Cammalleri, M.; Castellano, A. Analysis of hybrid vehicle transmissions with any number of modes and planetary gearing: Kinematics, power flows, mechanical power losses. *Mech. Mach. Theory* **2021**, *162*, 104350. [CrossRef]
4. Chan, C.; Yang, D.; Volz, T.; Breitweiser, D.; Jamzadeh, F.S.; Frank, A.A.; Omitsu, T. System Design and Control Considerations of Automotive Continuously Variable Transmissions. *Am. Soc. Mech. Eng.* **1984**, 840048.
5. Fox, A.J. *Design and Analysis of A Modified Power-Split Continuously Variable Transmission*; West Virginia University: Morgantown, WV, USA, 2003.
6. Vahabzadeh, H.; Macey, J.P.; Dittrich, O. A Split-Torque, Geared-Neutral Infinitely Variable Transmission Mechanism. *SAE Tech. Pap.* **1990**, 905089.
7. Beccari, A.; Cammalleri, M. Implicit regulation for automotive variators. *Proc. Inst. Mech. Eng. Part D J. Automob. Eng.* **2001**, *215*, 697–708. [CrossRef]
8. Perry, F.G. The Perbury Transmission. *Am. Soc. Mech. Eng.* **1980**, *80*, 800302.
9. Miller, J.M.; Everett, M. An assessment of ultra-capacitors as the power cache in Toyota THS-II, GM-allision AHS-2 and Ford FHS hybrid propulsion systems. In Proceedings of the IEEE Applied Power Electronics Conference and Exposition, Austin, TX, USA, 6–10 March 2005; pp. 481–490.
10. Chan, C.C. The state of the art of electric and hybrid vehicles. *Proc. IEEE* **2002**, *90*, 247–275. [CrossRef]
11. Volpe, S.S.; Carbone, G.; Napolitano, M.; Sedoni, E. Design Optimization of Input and Output Coupled Power Split Infinitely Variable Transmissions. *J. Mech. Des.* **2009**, *131*, 111002. [CrossRef]
12. Kim, J.; Kim, T.; Min, B.; Hwang, S.; Kim, H. Mode Control Strategy for a Two-Mode Hybrid Electric Vehicle Using Electrically Variable Transmission (EVT) and Fixed-Gear Mode. *IEEE Trans. Veh. Technol.* **2011**, *60*, 793–803. [CrossRef]
13. Zhang, F.; Yang, F.; Xue, D.; Cai, Y. Optimization of compound power split configurations in PHEV bus for fuel consumption and battery degradation decreasing. *Energy* **2018**, *169*, 937–957. [CrossRef]
14. Rotella, D.; Cammalleri, M. Power losses in power-split CVTs: A fast black-box approximate method. *Mech. Mach. Theory* **2018**, *128*, 528–543. [CrossRef]
15. Rotella, D.; Cammalleri, M. Direct analysis of power-split CVTs: A unified method. *Mech. Mach. Theory* **2018**, *121*, 116–127. [CrossRef]
16. Cammalleri, M.; Rotella, D. Functional design of power-split CVTs: An uncoupled hierarchical optimized model. *Mech. Mach. Theory* **2017**, *116*, 294–309. [CrossRef]
17. Nitz, L.; Truckenbrodt, A.; Epple, W. The new two-mode hybrid system from the global hybrid cooperation. In Proceedings of the 27th International Vienna Engine Symposium, Rio de Janeiro, Brazil, 25–27 October 2006; pp. 201–215.
18. De Pinto, S.; Mantriota, G. A simple model for compound split transmissions. *Proc. Inst. Mech. Eng. Part D J. Automob. Eng.* **2014**, *228*, 549–564. [CrossRef]
19. Bottiglione, F.; Mantriota, G. Power Flows and Efficiency of Output Compound e-CVT. *Int. J. Veh. Technol.* **2015**, *2015*, 1–12. [CrossRef]
20. de Pinto, S.; Mantriota, G. Power Flows in Compound Transmissions for Hybrid Vehicles. *Machines* **2019**, *7*, 19. [CrossRef]
21. Muller, H.W.; Mannhardt, W.G.; Glover, J.H. *Epicyclic Drive Trains Analysis, Synthesis, and Applications*; Wayne State University Press: Detroit, MI, USA, 1982.
22. Lulhe, A.M.; Date, T.N. A technology review paper for drives used in electrical vehicle (EV) & hybrid electrical vehicles (HEV). In Proceedings of the 2015 International Conference on Control Instrumentation Communication and Computational Technologies, Kumaracoil, India, 18–19 December 2015; pp. 632–636.

applied sciences

MDPI

Article

On the Use of a Hydrogen-Fueled Engine in a Hybrid Electric Vehicle

Stefano Beccari

Department of Engineering, University of Palermo, Viale delle Scienze, 90128 Palermo, Italy; stefano.beccari@unipa.it

Featured Application: The specific application of this research could be the implementation of a hydrogen-fueled engine in a hybrid electric vehicle in order to improve the engine efficiency and to eliminate its exhaust emissions.

Abstract: Hybrid electric vehicles are currently one of the most effective ways to increase the efficiency and reduce the pollutant emissions of internal combustion engines. Green hydrogen, produced with renewable energies, is an excellent alternative to fossil fuels in order to drastically reduce engine pollutant emissions. In this work, the author proposes the implementation of a hydrogen-fueled engine in a hybrid vehicle; the investigated hybrid powertrain is the power-split type in which the engine, two electric motor/generators and the drive shaft are coupled together by a planetary gear set; this arrangement allows the engine to operate independently from the wheels and, thus, to exploit the best efficiency operating points. A set of numeric simulations were performed in order to compare the gasoline-fueled engine with the hydrogen-fueled one in terms of the thermal efficiency and total energy consumed during a driving cycle. The simulation results show a mean engine efficiency increase of around 17% when fueled with hydrogen with respect to gasoline and an energy consumption reduction of around 15% in a driving cycle.

Keywords: hydrogen; hybrid electric vehicle; pollutant emissions

Citation: Beccari, S. On the Use of a Hydrogen-Fueled Engine in a Hybrid Electric Vehicle. *Appl. Sci.* **2022**, *12*, 12749. https://doi.org/10.3390/app122412749

Academic Editor: Adrian Irimescu

Received: 24 October 2022
Accepted: 7 December 2022
Published: 12 December 2022

1. Introduction

Road vehicles, both heavy and light duty, are primarily responsible for the pollutant and greenhouse gas emissions of the transportation sector. In recent decades, great research efforts have been made to reduce the environmental impact of road vehicles following two main strategies: the implementation of non-fossil fuels (bio-fuels, green hydrogen and electricity from renewable sources) and overall vehicle efficiency increases. Hybrid Electric Vehicles (HEV) are an example of the second strategy since they implement an Internal Combustion Engine (ICE), which is the main energy source, coupled with one or more electric machines and a battery pack that are meant to support the ICE during transient operations and to recover the vehicle's kinetic energy during deceleration phases.

The power-split powertrain configuration is one of the most economic and efficient; it is endowed of a planetary gear set (the power split device) that couples together the ICE, two electric motor/generators and the driveshaft connected to the wheels. This particular arrangement allows to operate the engine independently from the wheels and then to exploit, for each road load condition, the best ICE efficiency operating condition [1,2]. As far as greenhouse gasses emission reductions are concerned, the most promising alternative to fossil fuels, as an energy carrier, is represented by the so-called green hydrogen, produced by water electrolysis with renewable energy sources [3,4]. A stoichiometric air-hydrogen mixture has a volumetric power density comparable with gaseous hydrocarbons, hydrogen can be stored and transported, and its combustion produces only water.

There is a huge amount of literature on the use of hydrogen as a fuel in ICE [5–8] exploring the pro and cons of its application; however, one undeniable conclusion is that the engine emissions are almost free from carbon dioxides (except the few coming from engine lubricant) and under fuel lean combustion operation, also free from nitrous monoxide [5].

Starting from the two important conclusions that hybridization improves the ICE engine efficiency and that the hydrogen-fueled ICE is the cleanest possible, the author explored the possibility to hybridize a hydrogen-fueled ICE in order to compare its efficiency with that of the same hybridized engine when fueled with gasoline; a series of numeric simulations, detailed in the subsequent section, was made with the purpose to compare the hydrogen-fueled ICE with the gasoline one in terms of the Brake Thermal Efficiency (BTE) and total energy consumption in a driving cycle (DC). The implementation of a hydrogen-fueled ICE in an HEV is not new, some examples can be found in the literature—both theoretical [9–11] and practical [12–14]—proving that the topic is attractive, and the present work fit perfectly into this trend.

In [11], a fuel consumption reduction of 12.6% was reported for a hydrogen-fueled engine, in an HEV, compared to its gasoline counterpart; in [12], a 14.32% reduction in DC fuel consumption was reported for a hydrogen-enriched diesel engine, in an HEV, with respect to the standard diesel engine equipped in a conventional vehicle. In [13], a 20% fuel consumption reduction was reported when comparing a hydrogen-fueled HEV with a gasoline-fueled conventional vehicle.

All the above-mentioned studies either refer to hydrogen-enriched conventional fuels or refer to existing conventional engines converted to hydrogen and implemented in HEV but are experimental studies; there are few pure theoretical studies proving the superiority of hydrogen over gasoline in an HEV based on engine fuel consumption maps as performed in this work. The results of the present theoretical study largely confirm the engine efficiency and fuel consumption improvements reported in the literature.

The work proposed here starts from an existing spark-ignition ICE whose behavior, in terms of performance efficiency and pollutant emissions, has been widely explored experimentally both in gasoline and in natural gas (NG) operating modes [15–21]. The engine has been simulated with a zero-dimensional (0-D) thermodynamic model [22,23] that, after proper calibration with experimental data [22], is able to predict the engine performance in terms of BTE or Brake Specific Fuel Consumption (BSFC) maps.

In the present work, a simulated BTE map of the mentioned engine fueled with hydrogen, as detailed in the following section, was obtained and compared with the BTE map of the gasoline-fueled engine in order to highlight the advantages of the hydrogen fuel in terms of engine BTE increases and DC energy consumption reduction. Three different comparisons were performed: at first the mean engine BTE, evaluated over all the possible engine operating conditions, was compared referring to a conventional vehicle application, then the mean engine BTE, evaluated only over the best efficiency operating points, was compared referring to an HEV that is able to, due to its power-split arrangement, exploit the ICE engine only in its best efficiency operating points.

Finally, the DC energy consumption of the hydrogen hybrid vehicle was compared with both the gasoline conventional and the gasoline hybrid vehicle. The results of all comparisons are in favor of the hydrogen-fueled engine, which proves to be the cleanest and most efficient candidate to substitute, in the near future, fossil fuels as an ICE power source in both conventional and hybrid vehicles.

2. Numerical Simulations

A 0-D thermodynamic model, described in [22], properly calibrated using experimental data coming from a spark-ignition engine fueled with both gasoline and gaseous fuels [22,23], was used in the present work to obtain BTE maps of the engine when fueled with both gasoline and hydrogen. Table 1 shows the ICE specifications.

Table 1. Internal Combustion Engine (ICE) specifications.

Engine Specification	Value
brand and model	FIAT-FIRE 1.2 8v
n. of cylinders	4
Bore	70.8 (mm)
Stroke	78.86 (mm)
Rod to crank ratio	3.27
Compression ratio	9.8
Engine displacement	1242 (cm^3)
Engine maximum power	43.1 (kW)

The main model inputs are the engine speed, engine load (Manifold Absolute Pressure or MAP), air/fuel mass ratio and spark advance (SA); the main model outputs are the engine indicated and brake mean effective pressures (IMEP and BMEP), the engine indicated and brake thermal efficiencies (ITE and BTE) and the engine torque and power. With respect to the previous simulation version [22], which was able to evaluate only the in-cylinder pressure during the whole engine cycle and then the IMEP, an engine friction model was added in order to evaluate the friction mean effective pressure (FMEP) and, in turn, both the BMEP and BTE. The Chen and Flynn friction model [24] was implemented; according to this model, the engine FMEP can be evaluated using the following equation:

$$\mathrm{FMEP} = \mathrm{A} + \mathrm{B} \cdot p_{\max} + \mathrm{C} \cdot n + \mathrm{D} \cdot n^2 \quad (1)$$

where $p_{\max}$ is the maximum in-cylinder pressure; n is the engine speed; and A, B, C and D are calibration coefficients that were tuned by means of experimental data in [15].

To identify the air/fuel (A/F) ratio of a mixture, the coefficient of excess air λ is defined, it is the fraction between the actual A/F ratio and the stoichiometric one. In order to avoid dangerous combustion phenomena, such as pre-ignition and knocking, the hydrogen-fueled ICE must be operated with a lean A/F mixture [5–8] (i.e., with $\lambda > 1$), and this (together with the lower specific power of hydrogen) reduces the engine performance with respect to the gasoline operation mode; to fill this gap, one of the most used techniques is supercharging. In the literature, many combinations of supercharging pressure and λ values can be found that lead to knock-safe operation and allow obtaining the same performance as gasoline-fueled engine [5,25–27].

The only pollutant emitted by a hydrogen-fueled engine, with a stoichiometric A/F ratio, is nitrous oxides (NOx); however, as soon as the A/F mixture is doubled (i.e., $\lambda = 2$) a drastic drop in NOx emissions is found [5,25,26]. Even supercharged engines do not emit NOx as long as the maximum combustion temperature remains below 1800 K [5], and in all the simulations presented in this work, the maximum combustion temperature remains below 1700 K.

Resuming the literature findings, one can roughly say that, with $\lambda = 2$ and a supercharging pressure of 2 bar absolute, the hydrogen-fueled engine is able to restore the gasoline engine performances and almost eliminate the NOx emissions. Regarding the combustion speed, hydrogen exhibits a laminar flame velocity one order of magnitude greater than gasoline for the stoichiometric A/F ratio and around 60% greater than gasoline for $\lambda = 2$ [27,28]. Considering the same engine and operating conditions, the turbulence inside the combustion chamber should be the same for both gasoline and hydrogen fuels, and then the ratio between the laminar flame velocities of the two fuels can be roughly considered the same between the turbulent flame velocities.

This conclusion was confirmed by some preliminary experimental tests performed by the author on an ICE test bench. Resuming all the above-mentioned considerations, in order to simulate a hydrogen-fueled engine that should produce the same performance of the gasoline counterpart (i.e., the maximum power), that should not produce abnormal combustion phenomena and that should not emit NOx, the combustion duration must be set to roughly 60% of the gasoline counterpart, the A/F ratio must be set to $\lambda = 2$, and the engine must be supercharged in order to obtain MAP = 2 bar absolute.

The engine model was then upgraded in order to implement supercharging (SC) in two different ways: using a turbocharger (TC) driven by an exhaust gas turbine or using a volumetric compressor (VC) driven directly by the engine. In the first case, the engine MAP is set to a desired value that, for the sake of simplicity, was kept constant as the engine speed varied; the engine back pressure (p_b) was evaluated supposing that the same mass flows through both the compressor and turbine and equating the two specific works with the following equation:

$$p_b = \left(1 - \frac{T_{MAN}\left(\mathrm{MAP}^{\frac{k-1}{k}} - 1\right)}{\eta_c \eta_t T_{EXMAN}}\right)^{\frac{-k}{k-1}} \tag{2}$$

where k is the isentropic coefficient (1.4 for air); T_{MAN} is the inlet manifold air temperature that is set to the ambient temperature (300 K) considering the presence of an intercooler; η_c and η_t are the compressor and turbine isentropic efficiencies set to 0.85 and 0.9, respectively; and T_{EXMAN} is the exhaust manifold gas temperature set to a first-try value and then verified by simulations with an iterative process.

In the case of volumetric compressor supercharging (VCSC), as far as the compressor is driven directly by the engine, the corresponding FMEPc must be evaluated in order to obtain: BMEP = IMEP − FMEP − FMEPc. Assuming, for the sake of simplicity, an adiabatic screw compressor, the same volumetric efficiency for both engine and compressor and the presence of an intercooler between compressor and engine, one obtains:

$$\mathrm{FMEP_c} = \frac{k}{k-1}\mathrm{MAP}\left(\mathrm{MAP}^{\frac{k-1}{k}} - 1\right) \tag{3}$$

The desired MAP value depends on the compressor displacement V_c and rotating speed n_c:

$$\mathrm{MAP} = \frac{2V_c n_c}{Vn} \tag{4}$$

where V and n are the engine displacement and rotating speed, respectively.

Three engine configurations were simulated: the naturally aspirated gasoline-fueled engine, the turbocharged (TC) hydrogen-fueled engine and the supercharged (SC), with volumetric compressor (VC) hydrogen-fueled engine. For all the configurations, the engine speed and MAP were varied obtaining many different operating conditions. For each operating condition, λ and SA were set as the optimal values (i.e., the maximum brake torque SA) or the knock-limited values (in the case of gasoline-fueled engines, those values were experimentally pre-determined [15–17]).

Some preliminary simulations were performed in order to find the MAP value of the boosted hydrogen-fueled engine able to produce the same maximum power of the gasoline version (43.1 kW): for the TC version, this value was MAP = 2.1 bar absolute (in accordance with the above-mentioned considerations) while, for the VCSC version, the value was MAP = 2.7 bar absolute; the second MAP value was higher because of the higher energy needed to drive the compressor.

Figure 1 shows the λ and SA maps used in the gasoline-fueled engine simulation for all the operating conditions. The SA value is expressed in crank angle degrees (CAD) before top dead center (BTDC).

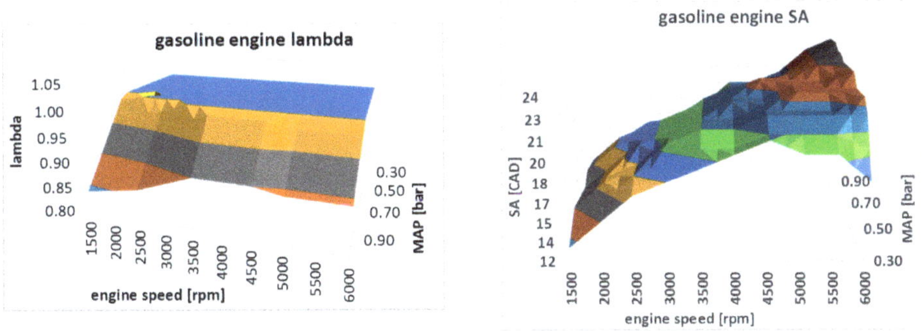

(**a**) (**b**)

Figure 1. λ (**a**) and SA (**b**) maps of the gasoline-fueled engine vs. engine speed [rpm] and MAP [bar].

For both hydrogen configurations, the A/F ratio is fixed at $\lambda = 2$ for all the operating conditions, while the SA is set to the maximum brake torque value. Figure 2a,b shows the SA maps adopted for the TC and SC engine, respectively.

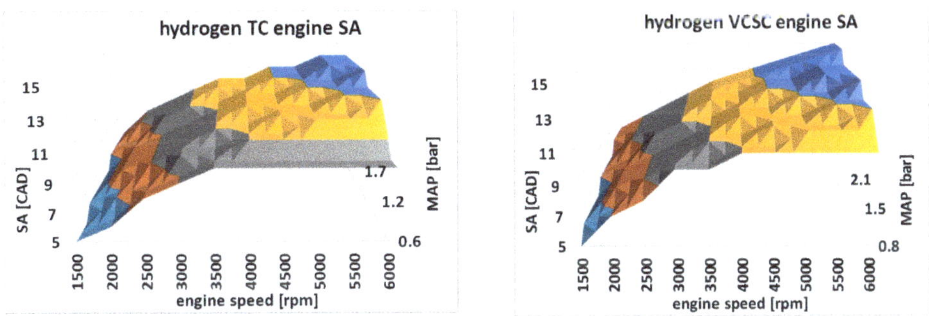

(**a**) (**b**)

Figure 2. SA maps of the TC (**a**) and VCSC (**b**) hydrogen-fueled engine vs. engine speed and MAP.

The main differences between gasoline- and hydrogen-fueled engine simulations are that the gasoline engine is operated with a rich A/F mixture and limited SA near full load conditions (Figure 1) to avoid knocking phenomena, and this lowers the engine BTE compared to hydrogen configurations; and the combustion duration of a gasoline-fueled engine is set to around 80 CAD [16,22,23] while the hydrogen combustion duration is set to around 50 CAD (roughly 60% lower than gasoline as stated above), and this further increases the hydrogen engine BTE compared to gasoline operation.

As already stated before, the hybridization obtains the best results for a conventional engine, and the scope of this work is to compare the hydrogen- and gasoline-fueled engines on their best operating conditions. The ICE is supposed to be implemented in a series-parallel hybrid vehicle equipped with a power-split device and two electric motor-generators in order to obtain the advantages of both series and parallel configurations as well as a high voltage battery pack (140 V) able to both recover the kinetic energy of the decelerating vehicle and drive the electric motor during transient operations.

The electric motors maximum power can be set to around 80% of the ICE maximum power (36 kW), and the battery pack energy capacity can be set to around 900 Wh. The above-described hybrid vehicle specifications correspond to that of a light duty passenger car [29].

In order to test the three hybridized engines (gasoline, VCSC and TC hydrogen) in a realistic situation, the Worldwide Harmonized Light Vehicles Test Cycle (WLTC) was considered. The cycle is composed of 1800 speed and acceleration values divided by a one second time interval (a fine discretization). For each speed value of the WLTC, the power required by the vehicle (PRV) was evaluated as the product between the speed and the force required to push the vehicle at that speed, sum of the acceleration force F_a and the drag force F_d. Since both the speed and acceleration are known, for each operating point of the WLTC, it is possible to evaluate F_a with the following equation:

$$F_a = (m + m_r)a \tag{5}$$

where m is the vehicle curb weight, a is its acceleration and m_r is the total equivalent mass of the rotating elements.

F_d is the sum of aerodynamic and rolling resistance forces and can be evaluated with the following equation:

$$F_d = \frac{1}{2}A_d\rho v^2 + mgf_r \tag{6}$$

where ρ is the air density, v is the vehicle speed, A_d is the drag area (product of the vehicle frontal area and the drag coefficient), g is the gravitational acceleration and f_r is the wheels' rolling friction coefficient. Assuming m_r = 45 kg, m = 1000 kg, A_d = 0.75 m^2 (mean values of a standard passenger car [29]) and f_r = 0.01, the PRV was evaluated for each point of the WLTC. The power supplied by the engine (PSE) was evaluated with the following equation:

$$\text{PSE} = \frac{\text{PRV}}{\eta_t} \tag{7}$$

where η_t is the transmission efficiency for which a value of 0.9 can be set [30–33].

Assuming that in a split-power HEV the engine is always operated in its best efficiency points, for each WLTC operating point a specific procedure (described in the following section) was used to link a specific engine efficiency to each *PSE* and then to evaluate the power required by the engine (PRE); integrating the PRE values along the 1800 s of the WLTC duration, the total required energy was found.

Considering that an HEV is able to recover a part of the energy required to decelerate the vehicle, all the operating points of WLTC that involve a negative *PRV* were integrated, and the corresponding total energy, multiplied by an efficiency factor, was subtracted to the total required energy to obtain the total energy consumption. The efficiency factor, considering the conversion efficiency of the electrical machines and batteries involved, was set to a conservative value of 0.65.

To have a "low efficiency" reference, the conventional gasoline engine was tested with the above-described procedure; a five-speed gearbox with the following gear ratios was set: t_1 = 0.36, t_2 = 0.53, t_3 = 0.85, t_4 = 1.14 and t_5 = 1.22, and the final drive ratio was 0.29. For each WLTC vehicle speed, the engine speed was evaluated considering the following gear change speed for each gear ratio: v_1 = 35 km/h, v_2 = 70 km/h, v_3 = 112 km/h and v_4 = 168 km/h.

For each WLTC point, knowing the engine speed and the *PSE*, a specific engine BTE was taken from the simulated engine map in Figure 3a (shown in the following section) and used to evaluate the corresponding PRE; integrating all the PRE values over the WLTC time duration, the total required energy was evaluated and, due to the absence of energy recovery systems, coincides with the total energy consumption. Clearly, this procedure does not exploit the best engine efficiency points because the operating condition is bound by the engine speed that, in turn, is bound by the fixed engine-driveshaft gear ratio.

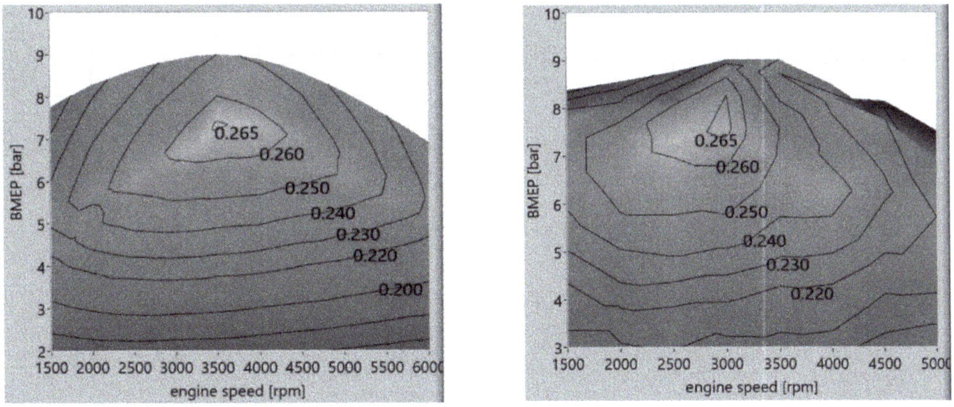

(**a**) (**b**)

Figure 3. Simulated (**a**) and experimental (**b**) BTE maps of the gasoline-fueled engine.

3. Results and Discussion

The results of gasoline- and hydrogen-fueled engine simulations are presented in terms of BTE instead of BSFC because of the different lower heating values of the two fuels. Figure 3 shows the simulated (a) and experimental (b) BTE maps of the gasoline engine as a function of engine speed and BMEP (proportional to engine torque). The maximum efficiency is 0.265, and it is located below the full load condition for the above-mentioned considerations.

Figure 4 shows the BTE maps coming from the hydrogen-fueled engine simulation with the TC (a) configuration and the VCSC (b) configuration; some relevant aspects can be highlighted: the maximum BTE of the SC engine (0.277) is higher than the gasoline one but lower than the TC one (0.305); furthermore, the maximum BTE zone of the SC engine is well below the full load condition, such as in the gasoline BTE map while, in the TC configuration, the maximum efficiency zone is located at maximum BMEP. The TC configuration performs better than the VCSC one likely due to high compression work at high MAP levels in the second case.

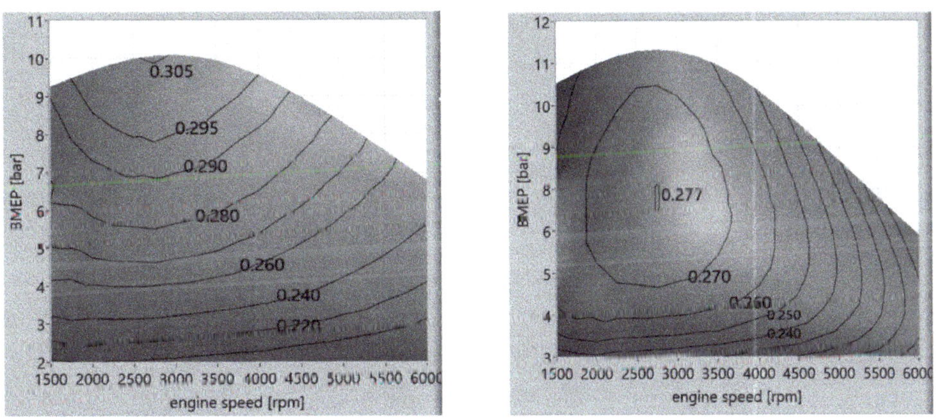

(**a**) (**b**)

Figure 4. BTE maps of the hydrogen-fueled engine.

When the engine equips a conventional vehicle, for a given road load (i.e., a required power), there are infinite possible combinations of engine speed and load (operating points) that produce that power; however, only one is the best efficiency point. This scenario is represented in Figure 5 where the gasoline-fueled engine BTE map is displayed together with two curves at constant required power (10 and 20 kW); the curves cross the BTE surface in many different points but only the two marked with a green cross are the best efficiency points (i.e., the points where the constant power curve is tangent to a constant BTE curve).

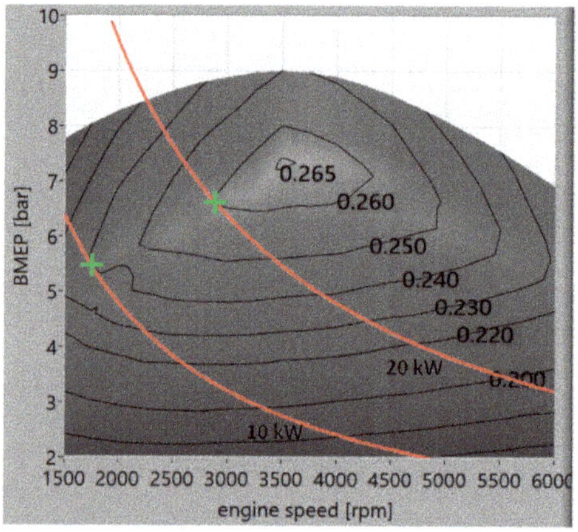

Figure 5. The best efficiency points at constant required power (gasoline).

The main difference between a conventional vehicle and a split-power hybrid one is that, in the first case, the ICE operating point depends on both the vehicle operating point and the current gear ratio between engine shaft and wheels while, in the second case, the engine can always be operated in the best efficiency points due to the decoupling between ICE and wheels and to the application of energy management control strategies (EMCS).

To find the mean BTE of the three engine configurations described above, equipping a conventional vehicle, one should evaluate the average BTE over all the points in the maps reported in Figures 3a and 4a,b because all the points are potential operating conditions.

Table 2 shows the mean BTE comparison for the three analyzed engine configurations: the same trend of the maximum BTEs is observed with the TC hydrogen-fueled engine that outperforms the gasoline operated one (almost 20% mean BTE increase), and the TC setup is slightly better than the VCSC one.

Table 2. The mean BTE of the three engine configurations in a conventional vehicle.

Engine Configuration	Mean BTE	% Increase Compared to Gasoline
gasoline	0.211	0
hydrogen VCSC	0.246	16.5%
hydrogen TC	0.253	19.5%

As already stated before, the hybridization obtains the best results from a conventional engine, and the scope of this work is to compare the hydrogen- and gasoline-fueled engines at their best operating conditions. Assuming that, theoretically, a split-power HEV is able to operate the engine in its higher BTE operating point for each required power, the best efficiency path (BEP) is defined as the curve in the BTE map that links all the maximum BTE points for each possible required power from zero to the maximum (43.1 kW).

Figure 6 shows the gasoline BTE map with different constant power curves and, in green, the BEP that links all the tangent points between a constant power curve and an iso-efficiency curve. The mean engine BTE, in this case, was evaluated as the average value over all the points lying in the BEP.

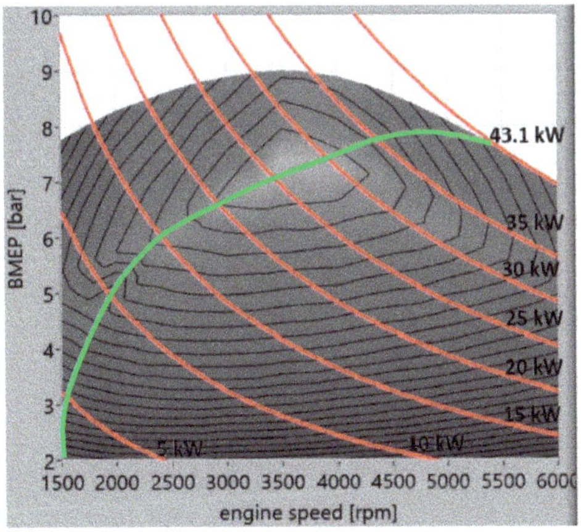

Figure 6. The gasoline-engine BTE map with the BEP in green.

Figure 7 shows the BTE maps of hydrogen-fueled engines in both TC and VCSC configuration with the BEP curves. The mean BTE value of the hydrogen-fueled engine equipped in an HEV is evaluated by averaging the BTE values lying on the BEP curves.

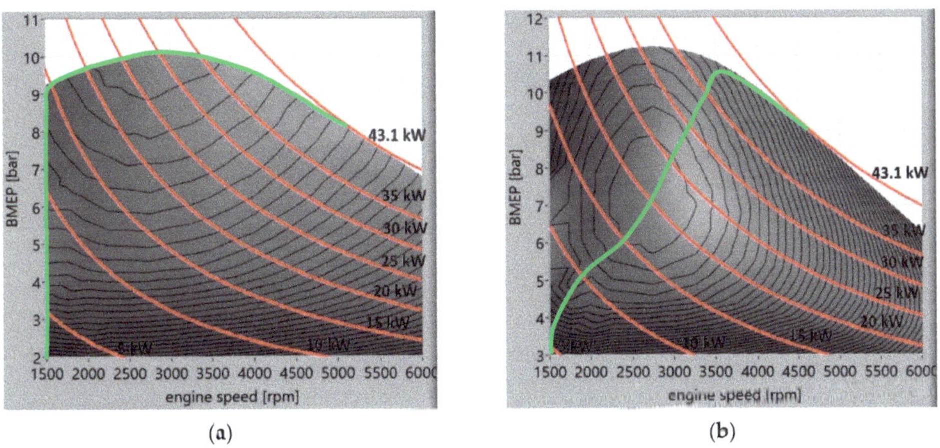

Figure 7. The hydrogen-engine BTE map with the BEP in green.

Table 3 shows, for all the three engine configurations, the power and BTE values of all the points lying on the BEP.

Table 3. The engine power and BTE of points lying on BEPs.

Power [kW] Gasoline	BTE Gasoline	Power [kW] SC Hydrogen	BTE SC Hydrogen	Power [kW] TC Hydrogen	BTE TC Hydrogen
3.9	0.182	4.2	0.226	3.6	0.212
4.6	0.195	5.4	0.247	4.5	0.232
5.4	0.206	6.4	0.258	5.5	0.248
6.9	0.223	7.4	0.263	6.3	0.259
7.6	0.230	8.3	0.266	7.2	0.266
8.4	0.233	9.2	0.267	8.0	0.271
9.2	0.236	10.1	0.268	8.8	0.276
10.1	0.241	10.9	0.268	9.6	0.280
11.0	0.238	11.9	0.268	10.3	0.279
11.9	0.247	12.7	0.272	11.1	0.282
12.4	0.245	13.9	0.272	11.9	0.285
13.6	0.251	14.5	0.274	12.7	0.287
14.9	0.249	15.9	0.275	13.6	0.289
15.5	0.253	16.4	0.276	14.4	0.291
16.9	0.253	17.2	0.275	15.2	0.292
17.4	0.255	17.9	0.276	16.2	0.294
18.9	0.258	19.4	0.276	17.2	0.296
19.2	0.257	20.9	0.276	18.9	0.298
20.9	0.262	21.6	0.277	20.0	0.300
22.0	0.250	21.8	0.276	21.6	0.301
22.8	0.263	23.6	0.276	22.9	0.302
24.1	0.257	24.8	0.275	24.4	0.302
25.5	0.255	25.3	0.275	25.8	0.304
26.6	0.265	26.3	0.274	27.0	0.303
27.8	0.261	27.0	0.274	28.7	0.305
28.4	0.264	28.7	0.273	29.6	0.302
29.1	0.258	29.1	0.272	31.3	0.304
30.1	0.256	30.3	0.271	31.9	0.301
31.4	0.261	31.8	0.269	33.8	0.303
32.6	0.258	32.6	0.268	34.1	0.300
33.8	0.257	33.3	0.267	36.0	0.302
34.6	0.254	34.4	0.266	37.4	0.294
36.1	0.252	35.7	0.265	38.0	0.299
37.5	0.250	36.1	0.264	38.6	0.290
38.5	0.247	37.7	0.261	39.6	0.296
39.9	0.245	39.2	0.259	40.8	0.292
41.0	0.241	40.8	0.254	41.8	0.288
41.9	0.237	42.0	0.249	42.5	0.283
42.6	0.234	42.7	0.243	42.9	0.277
43.1	0.226	43.1	0.236	43.1	0.271

Table 4 shows the average values extracted from Table 3, which are the mean BTE of a gasoline- or hydrogen-fueled ICE when equipped in an HEV. Some important considerations can be drawn: The ICE hybridization brings great benefits to the mean BTE in both gasoline (+15.9%) and hydrogen (+8.0% and +13.3%) configurations. TC remains, also in HEV, a better alternative than VCSC. Finally, the TC hydrogen ICE has a higher mean BTE than the gasoline hybridized engine in both the conventional and hybridized configurations with increases of +3.1% and +16.7%, respectively.

Table 4. The mean BTE of the three engine configurations in an HEV.

HEV ICE Configuration	Mean BTE	% Increase Compared to Gasoline Hybrid	% Increase Compared to Conventional
gasoline	0.245	0	15.9%
hydrogen VCSC	0.266	8.6%	8.0%
hydrogen TC	0.286	16.7%	13.3%

In order to test the three hybridized engines (gasoline, VCSC and TC hydrogen) in a more realistic situation, they were implemented in the WLTC driving cycle as described above. Table 5 resumes the findings, in terms of the total energy consumed in the WLTC, of both gasoline- and hydrogen-fueled engines. All the hybrid configurations show a strong reduction of energy consumption compared to the conventional gasoline engine, from the −15% of the hybrid gasoline to the −23.6% of the VCSC hybrid and finally to the −28.2% of the TC hybrid, and this confirms the effectiveness of the hybrid technology in increasing the vehicle efficiency by exploiting the engine best efficiency operating points and by recovering the vehicle kinetic energy.

Table 5. WLTC energy and fuel mass consumption comparison.

ICE Configuration	Total Energy Consumption [kJ]	Energy Difference Compared to Gasoline Hybrid [%]	Energy Difference Compared to Gasoline Conventional [%]	Total Fuel Mass Consumption [kg]
gasoline conventional	50,731	+18.0%	0	1.153
gasoline hybrid	42,930	0	−15.0%	0.976
hydrogen VCSC hybrid	38,749	−9.7%	−23.6%	0.323
hydrogen TC hybrid	36,430	−15.1%	−28.2%	0.304

Table 5 highlights the efficiency increases specifically due to the use of hydrogen in a hybrid vehicle with respect to gasoline (third column). The hydrogen VCSC hybrid engine shows a total energy consumption that is 9.7% lower than the gasoline hybrid, and the hydrogen TC hybrid engine is 15.1% lower than the gasoline hybrid confirming the already discussed mean efficiency increases reported in Table 4.

The total fuel mass consumption of hydrogen (in both TC and VCSC configurations) was lower than one third that of gasoline, and the main reason is the lower heating value of hydrogen that is three times that of gasoline. In any case, a vehicle mileage evaluation can be made. The energy consumption reduction of 15.1% agrees well with the above-mentioned literature results [9–14] in which a range between 12% and 20% of gasoline equivalent fuel consumption reduction has been reported between conventional fuel and hydrogen in HEV.

Considering that a 50 L cylinder at 300 bar pressure contains 1.35 kg of hydrogen, and with 0.304 kg of hydrogen, a TC HEV can theoretically travel 23.262 km (WLTC equivalent travel distance), three cylinders would allow a 310.3 km travel distance, which is good mileage if compared with natural-gas-fueled light duty vehicles for which commercial data [34,35] and the scientific literature [36,37] report an average mileage of around 400 km considering a typical 13–14 kg fuel tank capacity.

To resume the findings, the hydrogen-fueled engine, also in its conventional configuration, performed better than the gasoline engine even if the latter was hybridized (Tables 2 and 4). If the hydrogen engine underwent hybridization, this performance gap increased even more (Table 4); this efficiency gap remained almost unchanged when analyzing the WLTC total energy consumption (Table 5), and a respectable mileage of more than 300 km was estimated for the TC hydrogen-fueled engine. This conclusion, together with the fact that the hydrogen-fueled engine is almost free from pollutant and greenhouse gasses emissions proves the superiority of green hydrogen over traditional fossil fuels in both conventional and hybrid vehicles.

4. Conclusions

This paper shows a theoretical comparison between a gasoline- and a hydrogen-fueled ICE equipped in an HEV; a set of simulations was performed in order to compare the two fuels in terms of both the engine efficiency and total energy consumption while keeping the maximum engine power constant. To obtain the same maximum power of the gasoline-fueled engine, the hydrogen engine must be supercharged, and the best solution proved to be a turbocharger driven by an exhaust gas turbine. The first comparison involves a conventional vehicle: the hydrogen turbocharged engine exhibited an average BTE 19.5% higher than that of the gasoline counterpart.

A second comparison between the two fuels was performed considering the engine aboard an HEV that exploits, for each operating condition, the best engine efficiency. In this case, the hydrogen TC engine showed an average BTE 16.7% higher than that of the gasoline one. Finally, the WLTC driving cycle was considered in order to evaluate the total energy consumption of the engine equipped in the HEV and fueled either by gasoline or by hydrogen. In this case, the total energy consumption of the hydrogen-fueled engine (HEV) was 15.1% lower than that of the gasoline one (HEV) and 28.2% lower if compared with the gasoline engine aboard a conventional vehicle.

An indicative mileage of 300 km was estimated for the hybrid hydrogen engine with a 150 L tank loaded at 300 bar pressure. The energy consumption reduction of 15.1% agrees with the above-mentioned literature results in which a range between 12% and 20% of gasoline equivalent fuel consumption reduction was reported between conventional fuel and hydrogen in HEVs.

The higher engine efficiency when fueled with hydrogen compared with gasoline is mainly due to three reasons: the lean mixture composition (while gasoline, at full load and works with rich mixture), the better combustion phasing allowed by the higher knocking resistance and the shorter combustion duration due to the higher flame speed. In this work, the three mentioned advantages of hydrogen were quantified in terms of attainable engine efficiency increase and energy consumption reduction in HEVs.

Our conclusion is that a hydrogen-fueled engine performs better than the gasoline counterpart in both conventional and hybrid vehicles; considering that hydrogen, under proper operating conditions, is a zero-emissions fuel, this represents an almost obligatory choice in the automotive field.

Funding: This research received no external funding.

Institutional Review Board Statement: Not applicable.

Informed Consent Statement: Not applicable.

Data Availability Statement: Not applicable.

Conflicts of Interest: The authors declare no conflict of interest.

References

1. Zhuang, W.; Li, S.; Zhang, X.; Kum, D.; Song, Z.; Yin, G.; Ju, F. A survey of powertrain configuration studies on hybrid electric vehicles. *Appl. Energy* **2020**, *262*, 114553. [CrossRef]
2. Singh, K.V.; Bansal, H.O.; Singh, D. A comprehensive review on hybrid electric vehicles: Architectures and components. *J. Mod. Transp.* **2019**, *27*, 77–107. [CrossRef]
3. Mazzeo, D.; Herdem, M.S.; Matera, N.; Wen, J.Z. Green hydrogen production: Analysis for different single or combined large-scale photovoltaic and wind renewable systems. *Renew. Energy* **2022**, *200*, 360–378. [CrossRef]
4. Lee, H.; Choe, B.; Lee, B.; Gu, J.; Cho, H.S.; Won, W.; Lim, H. Outlook of industrial-scale green hydrogen production via a hybrid system of alkaline water electrolysis and energy storage system based on seasonal solar radiation. *J. Clean. Prod.* **2022**, *377*, 134210. [CrossRef]
5. White, C.M.; Steeper, R.R.; Lutz, A.E. The hydrogen-fueled internal combustion engine: A technical review. *Int. J. Hydrog. Energy* **2006**, *31*, 1292–1305. [CrossRef]
6. Verhelst, S. Recent progress in the use of hydrogen as a fuel for internal combustion engines. *Int. J. Hydrog. Energy* **2014**, *39*, 1071–1085. [CrossRef]
7. Verhelst, S.; Wallner, T. Hydrogen-fueled internal combustion engines. *Prog. Energy Combust. Sci.* **2009**, *35*, 490–527. [CrossRef]
8. Saafi, M.A.; Ou, S.; Jiang, Y.; Li, H.; He, X.; Lin, Z.; Gan, Y.; Lu, Z.; Dawson, S.T.M. Exploring the potential of hydrogen in decarbonizing China's light-duty vehicle market. *Int. J. Hydrog. Energy* **2022**, *47*, 36355–36371. [CrossRef]
9. Keller, J.; Lutz, A. Hydrogen Fueled Engines in Hybrid Vehicles. *SAE Trans.* **2001**, *110*, 481–486.
10. Arat, H.T. Simulation of diesel hybrid electric vehicle containing hydrogen enriched CI engine. *Int. J. Hydrog. Energy* **2019**, *44*, 10139–10146. [CrossRef]

11. Wang, Y.; Biswas, A.; Rodriguez, R.; Keshavarz-Motamed, Z.; Emadi, A. Hybrid electric vehicle specific engines: State-of-the-art review. *Energy Rep.* **2022**, *8*, 832–851. [CrossRef]
12. Arat, H.T. Alternative fuelled hybrid electric vehicle (AF-HEV) with hydrogen enriched internal combustion engine. *Int. J. Hydrog. Energy* **2019**, *44*, 19005–19016. [CrossRef]
13. He, X.; Maxwell, T.; Parten, M.E. Development of a Hybrid Electric Vehicle with a Hydrogen-Fueled IC Engine. *IEEE Trans. Veh. Technol.* **2006**, *55*, 1693–1703. [CrossRef]
14. Nakajima, Y.; Yamane, K.; Shudo, T.; Hiruma, M.; Takagi, Y. Research and Development of a Hydrogen-Fueled Engine for Hybrid Electric Vehicles. *SAE Trans.* **2000**, *109*, 1175–1179.
15. Pipitone, E.; Beccari, S.; Genchi, G. Supercharging the Double-Fueled Spark Ignition Engine: Performance and Efficiency. *J. Eng. Gas. Turbines Power.* **2017**, *139*, 102809. [CrossRef]
16. Beccari, S.; Pipitone, E. Detailed Combustion Analysis of a Supercharged Double-Fueled Spark Ignition Engine. *SAE Int. J. Engines* **2022**, *15*, 499–513. [CrossRef]
17. Beccari, S.; Pipitone, E.; Genchi, G. Knock onset prediction of propane, gasoline and their mixtures in spark ignition engines. *J. Energy Inst.* **2016**, *89*, 104–114. [CrossRef]
18. Genchi, G.; Pipitone, E.; Beccari, S.; Piacentino, A. Knock resistance increase through the addition of natural gas or LPG to gasoline: An experimental study. *SAE Tech. Pap.* **2013**, 100. [CrossRef]
19. Cammalleri, M.; Pipitone, E.; Beccari, S.; Genchi, G. A mathematical model for the prediction of the injected mass diagram of a S.I. engine gas injector. *J. Mech. Sci. Technol.* **2013**, *27*, 3253–3265. [CrossRef]
20. Pipitone, E.; Beccari, S.; Genchi, G. A refined model for knock onset prediction in spark ignition engines fueled with mixtures of gasoline and propane. *J. Eng. Gas. Turbines Power.* **2015**, *137*, 111501. [CrossRef]
21. Beccari, S.; Pipitone, E.; Cammalleri, M.; Genchi, G. Model-based optimization of injection strategies for SI engine gas injectors. *J. Mech. Sci. Technol.* **2014**, *28*, 3311–3323. [CrossRef]
22. Beccari, S.; Pipitone, E. An Effective Method to Model the Combustion Process in Spark Ignition Engines. *SAE Int. J. Engines* **2023**, *16*, 1–15. [CrossRef]
23. Beccari, S.; Pipitone, E. A New Simple Function for Combustion and Cyclic Variation Modeling in Supercharged Spark Ignition Engines. *Energies* **2022**, *15*, 3796. [CrossRef]
24. Chen, S.; Flynn, P. Development of a Single Cylinder Compression Ignition Research Engine. *SAE Tech. Pap.* **1965**, *74*, 650733. [CrossRef]
25. Nagalingam, B.; Dübel, M.; Schmillen, K. Performance of the Supercharged Spark Ignition Hydrogen Engine. *SAE Tech. Pap.* **1983**, *559*, 831688. [CrossRef]
26. Berckmüller, M.; Rottengruber, H.; Eder, A.; Brehm, N.; Elsässer, G.; Müller-Alander, G.; Schwarz, C. Potentials of a Charged SI-Hydrogen Engine. *SAE Tech. Pap.* **2003**, *1*, 3210. [CrossRef]
27. Natkin, R.; Tang, X.; Boyer, B.; Oltmans, B.; Denlinger, A.; Heffel, J.W. Hydrogen IC Engine Boosting Performance and NOx Study. *SAE Tech. Pap.* **2003**, *1*, 631. [CrossRef]
28. Dahoe, A.E. Laminar burning velocities of hydrogen–air mixtures from closed vessel gas explosions. *J. Loss Prev. Process. Ind.* **2005**, *18*, 152–166. [CrossRef]
29. Orecchini, F.; Santiangeli, A.; Zuccari, F.; Ortenzi, F.; Genovese, A.; Spazzafumo, G.; Nardone, L. Energy consumption of a last generation full hybrid vehicle compared with a conventional vehicle in real drive conditions. *Energy Procedia* **2018**, *148*, 289–296. [CrossRef]
30. Irimescu, A.; Mihon, L.; Pădure, G. Automotive transmission efficiency measurement using a chassis dynamometer. *Int. J. Automot. Technol.* **2011**, *12*, 555–559. [CrossRef]
31. Cammalleri, M.; Castellano, A. Analysis of hybrid vehicle transmissions with any number of modes and planetary gearing: Kinematics, power flows, mechanical power losses. *Mech. Mach. Theory* **2021**, *162*, 104350. [CrossRef]
32. Castellano, A.; Cammalleri, M. Power Losses Minimization for Optimal Operating Maps in Power Split HEVs: A Case Study on the Chevrolet Volt. *Appl. Sci.* **2021**, *11*, 7779. [CrossRef]
33. Castellano, A.; Cammalleri, M. Optimal Operation of Power-Split Hybrid Electric Powertrain: Comparison between Two Performance Indices. *Int. J. Mech. Control* **2022**, *23*, 3–14.
34. Commercial Data of CNG-Fueled Vehicles. Available online: https://www.marutisuzuki.com/cng (accessed on 28 November 2022).
35. Technical and Sales Data of CNG-Fueled Vehicles. Available online: https://www.rushlane.com/maruti-cng-sales-10-lakh alto wagonr-12430029.html (accessed on 28 November 2022).
36. Hochhauser, A.M.; Koehl, W.J.; Benson, J.D.; Burns, V.R.; Knepper, J.C.; Leppard, W.R.; Painter, L.J.; Rapp, L.A.; Rippon, B.H.; Reuter, R.M.; et al. Comparison of CNG and Gasoline Vehicle Exhaust Emissions: Mass and Composition—The Auto/Oil Air Quality Improvement Research Program. *J. Fuels Lubr.* **1995**, *104*, 1772–1798.
37. Wu, D.Y.; Matthews, R.D.; Popova, E.T.; Mock, C. The Texas Project, Part 4—Final Results: Emissions and Fuel Economy of CNG and LPG Conversions of Light-Duty Vehicles. *J. Fuels Lubr.* **1998**, *107*, 928–950.

applied sciences

MDPI

Article

CO_2 Emission Analysis for Different Types of Electric Vehicles When Charged from Floating Solar Photovoltaic Systems

Abinands Ramshanker [1], Suprava Chakraborty [2,*], Devaraj Elangovan [2], Hossam Kotb [3,*], Kareem M. Aboras [3], Nimay Chandra Giri [4,5] and Ephraim Bonah Agyekum [6]

1 School of Electrical Engineering, Vellore Institute of Technology (VIT), Vellore 632014, Tamil Nadu, India
2 TIFAC-CORE, Vellore Institute of Technology (VIT), Vellore 632014, Tamil Nadu, India
3 Department of Electrical Power and Machines, Faculty of Engineering, Alexandria University, Alexandria 21544, Egypt
4 Department of Electronics and Communication Engineering, Centurion University of Technology and Management, Jatni 752050, Odisha, India
5 Centre for Renewable Energy and Environment, Centurion University of Technology and Management, Jatni 752050, Odisha, India
6 Department of Nuclear and Renewable Energy, Ural Federal University Named after the First President of Russia Boris Yeltsin, 19 Mira Street, 620002 Ekaterinburg, Russia
* Correspondence: suprava1008@gmail.com (S.C.); hossam.kotb@alexu.edu.eg (H.K.)

Abstract: Renewable energy and electric vehicle technology are the two pillars for achieving a sustainable future. Floating solar power plants use PV modules on water infrastructure to save the land and increase module efficiency. Furthermore, the reduction in evaporation saves water. Electric vehicles are one of the fastest-growing markets and the most successful technologies to combat the problem of energy and climate change. This research aims to construct a floating PV system on the lake of the Vellore Institute of Technology (VIT), to analyze electric vehicle performance and greenhouse gas (GHG) emissions when charged using the installed floating PV system. To address this, a 1.5 MWP floating PV system was simulated and analyzed using Helioscope software. When charged from the proposed floating PV plant, electric bikes, scooters, and cars saved CO_2 emissions. When charged from a floating PV, E-bike, E-scooter, and E-car Net CO_2 emissions became zero in 25.5, 12.1, and 7.7 months, respectively. After the aforementioned time periods, all three electric vehicle types were zero-emission vehicles. The required charge for all three types of vehicles (1,000,000 km) was analyzed using a floating PV system. E-bike, E-scooter, and E-car CO_2 emission savings were −8,516,000 g/kWh, −328,000 g/kWh, and 525,600,000 g/kWh, respectively. All three types of electric vehicles can reduce CO_2 emissions for nations that rely on renewable energy, but only electric cars save carbon emissions over fixed distances. Through this research, we finally conclude that electric cars reduce CO_2 emissions the most compared to other electric vehicles.

Keywords: floating PV system; evaporation; e-mobility; electric car; CO_2 emission

Citation: Ramshanker, A.; Chakraborty, S.; Elangovan, D.; Kotb, H.; Aboras, K.M.; Giri, N.C.; Agyekum, E.B. CO_2 Emission Analysis for Different Types of Electric Vehicles When Charged from Floating Solar Photovoltaic Systems. *Appl. Sci.* **2022**, *12*, 12552. https://doi.org/10.3390/app122412552

Academic Editors: Marco Cammalleri, Vincenzo Di Dio and Antonella Castellano

Received: 24 October 2022
Accepted: 28 November 2022
Published: 7 December 2022

1. Introduction

Climate change has led to habitable places turning into deserts as wildfires and heatwaves occur increasingly around the world [1]. The warming of the Arctic has resulted in the melting of permafrost, glacial retreat, and sea ice loss [2]. It has affected the environment in very extreme ways, from sea levels rising to species going extinct. Climate change has been referred to as one of the most dangerous threats to human health by the World Health Organization (WHO) [3]. One of the best ways to mitigate the negative effects of climate change is to reduce the use of internal combustion engine-based vehicles and by adopting electric vehicles (EVs). Along with that, another way to mitigate the effects of climate change is by adopting renewable energy sources. The world target regarding electricity generated per year should be wholly from renewables compared to 25% (7000 TWh/year) at present. Among the various renewable energy sources available, solar energy has been getting much attention. The total installed capacity of solar energy at present is 384 GW and by 2050, the installation target is 8519 GW [4].

In recent years, a particular technology has been developed which involves fixing PV systems on canals, conduits, and channels. This technology is popularly called the floating solar PV system (FSPVS). The first proper FSPVS system was installed in 2007 in Aichi, Japan. It slowly began to increase with the yearly installed capacity reaching up to 1096 MWp per year. By the end of August 2020, the total global installed capacity of FSPVS would have reached 2.6 GWp. However, owing to COVID-19, installations were expected to decelerate in 2020, as seen in Figure 1 [5–8]. According to the report presented in [9], the worldwide floating solar market was valued at USD 2.55 billion in 2021 and is predicted to increase to USD 10.09 billion by 2030, with a compound annual growth rate (CAGR) of 16.5% between 2022 and 2030. This can be easily inferred from Figure 2. The number of research papers on floating PV systems in the last 15 years is shown in Figure 3. It is clear that FSPVS is an exponentially growing technology.

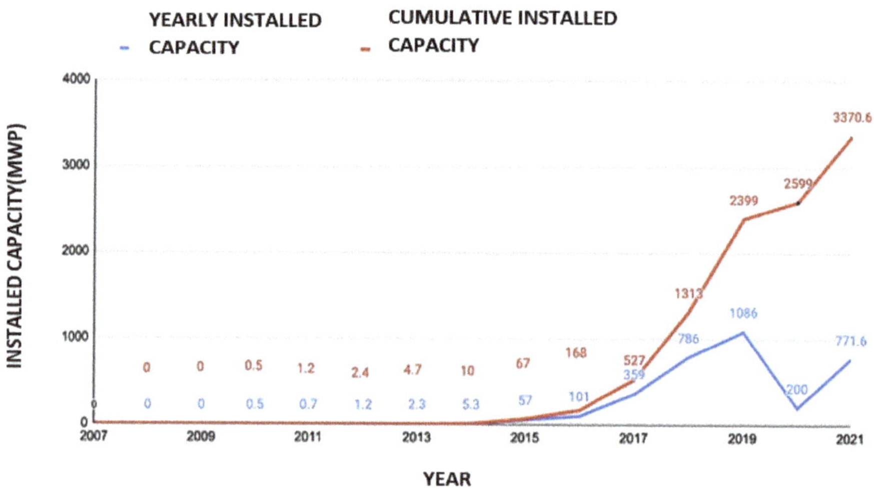

Figure 1. Yearly installed capacity of floating PV plant.

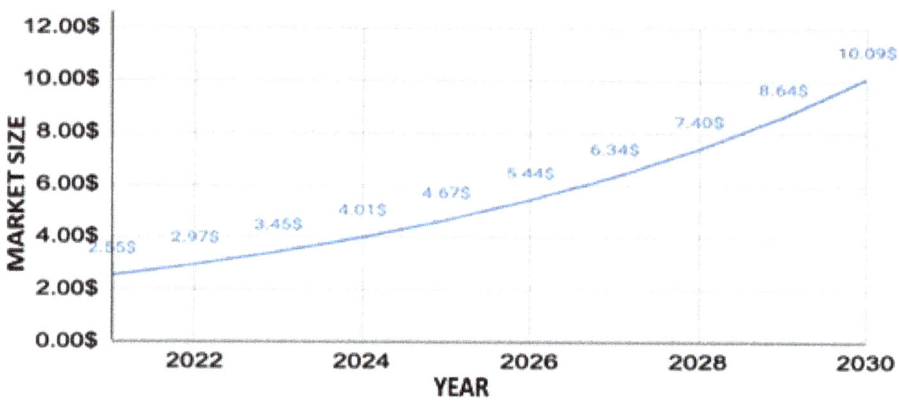

Figure 2. Total projected market size of floating PV [9].

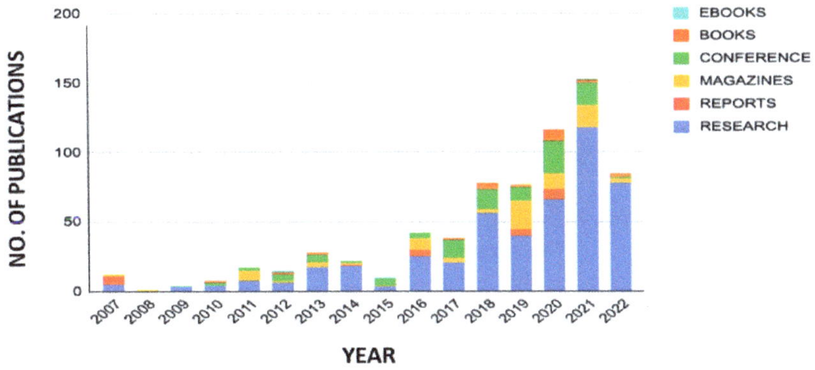

Figure 3. Number of publications based on floating PV systems by year.

Due to favorable government laws and the benefits of zero land usage and reduced evaporation, the deployment of PV systems on water bodies is quickly rising. The FSPVS system consists of the floating system, the mooring system, the PV system, the wires, and the connections.

(a) Floating system: The floating system is a combination of structure and floater. The PV system is mounted on top of the floating system.
(b) Mooring system: A mooring system is any structure to which a vessel may be anchored using cables or anchors. Mooring prevents FSPVS components from spinning or floating away.
(c) PV system: A PV system consists of PV modules and other power conditioning equipment for converting solar energy to electrical energy. In general, crystalline solar PV modules are utilized, although there is ongoing research into the type of PV modules installed in FSPVS. Sahu et al. [10] recently conducted an energy analysis of thin-film PV technology on land, water, and submerged systems, demonstrating the advantages of submerged installations.

Goswami et al. [11] conducted a thorough analysis of the various designs and architectures of floating FSPVS, as well as their economic implications. Theoretically, Sairam and Oliveria [12] have suggested numerous creative concepts for improving the performance of canal-top PV systems. The practical usefulness of these ideas, on the other hand, must be investigated. The influence of PV cover on water quality, water body ecology, and evaporation has been examined by Taye et al. [13]. Water bodies with excellent covering may have less of an influence on their environment. Figure 4 presents the structure and components of a PSPVS.

Many FPV projects have been researched in recent years. The most impactful ones have been the high-capacity projects. In [14], a study of a 10 MWp FPV plant in Bakreswar found that a floating solar PV plant has 10.2% more capacity than a land-based system. In [15], 13 Kwp, 45.1 MWp, and 163 MWp FPV were analyzed to show that they are cost-effective. In Mettur Dam, Tamil Nadu, a 150 MW FPV system with tracking on the dam of the hydro plant was analyzed and observed that it can reduce CO_2 emissions by 135.92 ktonnes [16]. Recently, in [17], a 6513 MW FPV was installed in Uttar Pradesh's Rajghat Dam to determine annual evaporation loss (or 0.9 L per kWh). The levelized cost of energy (LCOE) is USD 0.036/kWh (INR 2.61/kWh) with an IRR of 8.55%, which is encouraging for widespread FSPV deployment. This was the project with the highest installed capacity in 2022.

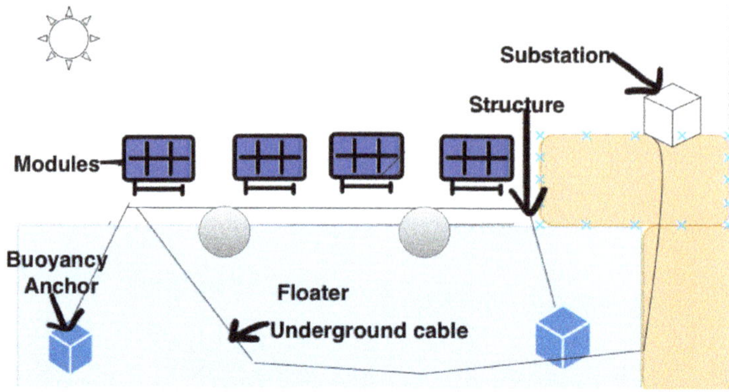

Figure 4. Components of a FSPVS.

Along with advancements in renewable energy, EV has been the biggest change that is being made in the world to lead a sustainable and clean future. The global electric sales have gone up exponentially. It began with one million electric vehicles sold globally in 2015 and increased exponentially up to ten million in 2021 [18], China and Europe seem to be the leaders in the EV market followed by Europe. The first milestone we expect to reach is 20 million EVs by 2025. After that, low estimates mention 33 million EVs by 2030, whereas high estimates talk about 47 million EVs [19]. Figure 5 represents the EV market Size from 2010–2020 and the projected market size till 2030. The low indicates the case where sustainable development policies are not adopted while the high estimate of 47 million refers to the case when sustainable development is followed. This provides the motivation to explore the performance analysis of EVs when charged by a FSPVS.

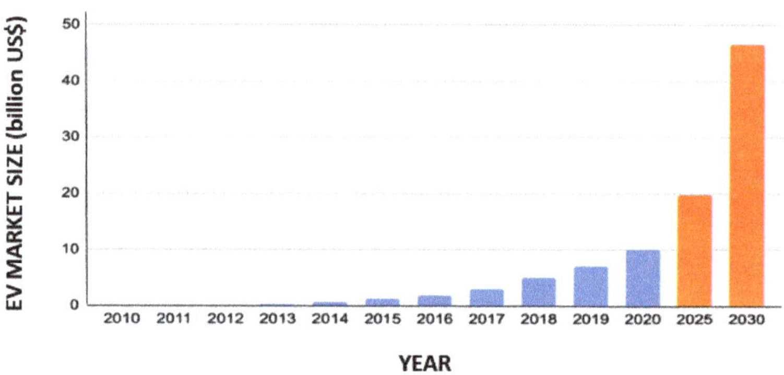

Figure 5. EV market size (in billion USD) from 2010–2020 and projected market size till 2030.

A research study was conducted by the Department of Energy [20] in which the complete cradle-to-grave lifecycle GHG emissions of EVs were compared to the average gasoline vehicle, which reached the same results as in [19], that hybrid EVs (HEVs), PEVs, and shorter-range BEVs cut greenhouse gas emissions, whereas longer-range BEVs are equivalent to the typical gasoline vehicles. Michalek et al. [21] quantified the lifetime GHG emissions and criteria pollutants of electric cars as well as the benefits of oil substitution. They discovered that EVs with larger battery packs are more costly, heavier, and emit more emissions than HEVs and PEVs with smaller battery packs, and that they give less emission gains. In both the basic and optimistic cases, all EVs cut emissions when compared to the average gasoline car. Shen [22] discovered that the manufacture of the EV accounts for 18.5% of the energy and 17% of the greenhouse gas emissions, while consumption accounted for 81.5% and 83%, respectively. Each study that analyzes the whole cradle-to-grave energy and greenhouse gas (GHG) emissions of electric vehicles concludes that emissions from vehicle usage surpass emissions from vehicle production.

The literature highlights the need to drive towards a more sustainable future by reducing the CO_2 emissions in EVs. This led to the motivation to utilize FSPVS to perform CO_2 analysis of EVs. This has not yet been explored in the literature.

1.1. Contribution

At present, the literature on Floating PV systems focuses mostly on installation and design approaches, but there has been nothing in the literature examining the system's benefits when charging electric vehicles. The implementation of EVs cannot help greatly towards the reduction in CO_2 emissions until or unless they are charged from a green grid instead of a conventional thermal grid. In this paper, a FSPVS of 1.5 MWp capacity has been designed and simulated at the natural lake of VIT, Vellore. The generated power has been analyzed and used to charge the E-bicycle, E-scooter and E-car to calculate the zero CO_2 emission periods of the said vehicles when life cycle CO_2 emission from each vehicle has been considered. The CO_2 emission profile of the said vehicles considering the fixed distance covered and charged from the FSPVS also has been analyzed. The FSPVS is further used to quantify the reduction in the rate of evaporation for a 1-year period. The objective of the present study is to:

- Design a floating system for VIT Lake, VIT, Tiruvalam Road, Katpadi, Vellore, Tamil Nadu, India;
- To analyze the annual energy generated from the FSPVS;
- Analyze the savings in water due to the prevention of evaporation by the FSPVS;
- Analyze and compare the performance and impacts on the environment while charging different electric vehicles by taking factors such as CO_2 emissions into consideration.

1.2. Structure of the Paper

The current paper has been organized as follows: after the Introduction, the Material and Methods behind this work is explained, in Section 2. Section 3 focuses on the results obtained and discussions related to them. The Results section focuses on the results obtained with the floating plant analysis and charging of EV using a floating solar PV system. Lastly, the work concludes in Section 4.

2. Materials and Methods

A simulation was made using Helioscope software to analyze the power generated when a floating PV system was employed at the suggested site location and the greenhouse gases emitted by different types of electric vehicles were compared. The data obtained from the simulation of the floating PV system were obtained and filtered. These data were then used as a reference for the case when different electric vehicles are charged. The setup was further used to quantify the reduction in the rate of evaporation for a 1-year period. We used a pre-existing analytical model to do this.

2.1. Case Study Area

The FSPVS of 1.5 MWp was proposed at the VIT Lake. The VIT Lake is the lake which is located inside the Vellore Institute of Technology University at Vellore in India. The VIT Lake is the largest lake at the Tiruvalam Road Area, located at 12.9723° N latitude and 79.1596° E longitude, respectively. The climatic parameters served as the input to our research.

2.2. Resource Assessment

The average temperature at Vellore was obtained for every month. The data regarding the average temperature were collected from [23]. The average temperature at Vellore was

22.6 °C during the month of January. It increased and peaked in May at a temperature of 31.2 °C. After that it decreased gradually until it became 22.5 °C. A similar trend was noticed for minimum temperature and maximum temperature. The detailed temperature data is presented below in Figure 6a. The average wind speed at Vellore followed a trend where it increased till July and then it decreased continuously till December, as shown in Figure 6a. In January the wind speed remained 7 mph. In March, there was a slight increase in the average wind speed to 7.5 mph. It increased further up to 10.6 mph. After July, the wind speed decreased to 6 mph which was the lowest of the year. After that it increased again up to 7.8 mph. The minimum and maximum solar potential increased till May and then decreased in June. It further increased till August and September before decreasing till December. This can be seen in Figure 6b. The performance of the solar PV system is affected by rainfall. Figure 6c describes the average rainfall and total number of rainy days in Vellore. We can see that on the month of October, the energy produced was lower than other months. This can be attributed to the total number or rainy days. The average sun hours remained low at 7.8 h in January but it then steadily increased and peaked in May at 10.6 h. It then decreased till December, becoming 6 h. This is described in Figure 6d.

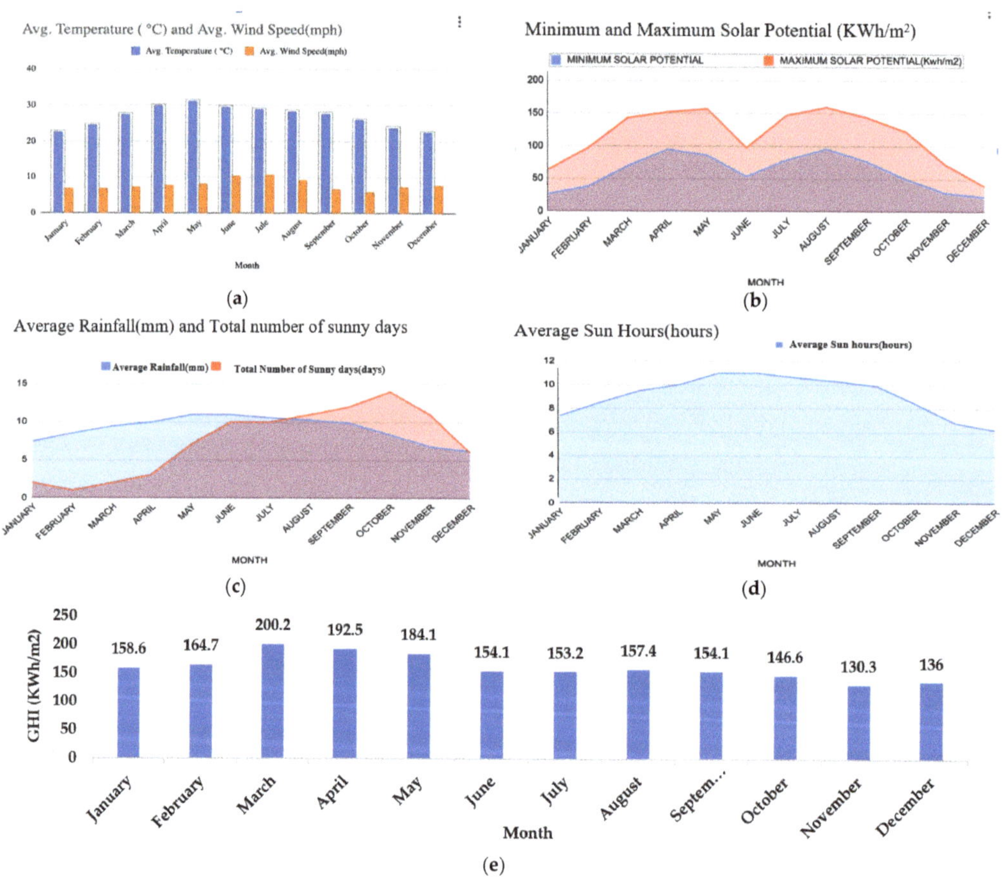

Figure 6. Monthly variation of various climatic parameters: (**a**) average temperature and wind speed; (**b**) minimum and maximum solar potential; (**c**) average rainfall and total number of sunny days; (**d**) average sun hours at Vellore; and (**e**) GHI irradiance per day by month.

The GHI irradiance/day for each month can be obtained from the research article [23]. Using the helioscope software, one can also obtain similar GHI values for the solar PV system. Figure 6e shows the GHI emissions obtained through the helioscope software.

2.3. System Design

4688 PV modules of Trina Solar TSM-PD14 3209 (May 15) [24] were chosen for installation. The STC Rating of the module was 320 W. The PV modules were connected to copper switches which were in turn connected to 50 nos. of SMA Sunny Tripower 24,000TL-US inverter [25]. The maximum input voltage of the inverter was 1000 V. The AWG Copper wires [26] were of string type and were 52,660 ft long. The component specifications are tabulated in Table 1.

Table 1. Specifications of module, inverter, and wire schedule.

Item name	Model	Parameter	Value
PV module	Trina Solar TSM-PD14 320	STC Rating	320 Wp
		Vmp	37.1 V
		Imp	8.63 A
		Voc	45.8 V
		Ioc	9.1 A
		Quantity	4689
Solar inverter	SMA Sunny Tripower 24,000 TL-US	Max AC Power Rating	24,000 W
		Max Input Voltage	1000 V
		Quantity	50
Wires	Wire schedule	Tier	String
		Wire	250 × 10 AWG
		Length	4958 ft

The PV modules were used to create a floating solar PV system. The panel consisted of a number of series and parallel connections of solar cells. The first panel was made up of two strings that represent the maximum number of PV modules that may be linked in series. In this string, 42 PV modules were used to give the proper voltage to the inverter. The voltage from the panel can be 1000 V as input of the inverter. The inverter was delivered in the 450 V to 800 V MPPT range. The inverter was connected to the external circuit and then this was then connected with the AC disconnector to protect the service panel from overvoltage and overcurrent follow.

2.4. EV Specifications

The specifications of the different vehicles are tabulated in Table 2 [27–29]

Table 2. Different electric vehicle parameters and CO_2 emission of electric vehicles.

Parameters	Battery Capacity (kWh)	Motor Power (kW)	Maximum Speed (km/h)	Maximum Torque (Nm)	Energy Consumption (Wh/km)	CO_2 Emission (g/psngr/km)
E-bike (BI I27)	0.28	0.25	25	40	7.9	16.1
E-scooter (Ather 450X)	2.61	6	80	26	30.7	29.8
E-car (Tesla Model 3)	50	336	162	639	151	92.4

2.5. Associated CO_2 Emissions Calculation for EVs

The energy consumed per km of different types of EVs was obtained from previously conducted studies and the CO_2 emissions were also obtained [30]. The CO_2 emission from the coal-based plant was 1000 g/kWh whereas the same was 40 g/kWh for the PV-based plant. The energy consumed per km for electric cars was 151 Wh, while it was 7.9 Wh for E-bikes and 30.7 Wh for E-scooters. From the analysis conducted using Helioscope, the annual energy generated from 1.5 MWp FSPVS was 2387.89 MWh. The total annual distance covered by the vehicle when charged from the generated energy of FSPVS is given by

$$annual\ distance\ covered\ in\ km\ (D_{k=1,2,3}) = \frac{EG}{EC_k} \tag{1}$$

where EG represents the annual energy generated from a 1.5 MW floating PV plant (Wh) and EC_k represents the energy consumption per km. The distance covered is represented by D_k, where $k = 1$ for the Bike, 2 for the Scooter and 3 for the car.

The percentage km covered more by the vehicle is given by:

$$D_{more,k} = \frac{\frac{EG}{ECk+1} - \frac{EG}{ECk}}{\frac{EG}{ECk+1}} \tag{2}$$

The annual CO_2 emission of the vechicle is given by the following equation

$$Annual\ CO_2\ emission,\ ACE_{k=1,2,3} = \frac{EG}{ECk} * CE_k \tag{3}$$

where CE_k represented the life cycle CO_2 emission/km from the vehicle. The saving in CO_2 emission per unit is seen as the difference between the CO_2 emission from the coal plant and the CO_2 emission from PV plant, where CEcp and CEpv represent the CO_2 emission from coal − based and PV − based plants, respectively.

$$Saving\ in\ CO_2\ emission\ per\ unit(S) = CEcp - CEpv \tag{4}$$

The total CO_2 emission savings (when charged from FPVS instead of the conventional grid) is given by the following equation,

$$Total\ CO_2\ emission\ savings(TCE) = EG * \frac{CEcp - CEpv}{1000} \tag{5}$$

Finally, the CO_2 emission saving when PV is used to charge the different types of EVs are calculated using

$$CES_{k=1,2,3} = TCE - ACE_{k=1,2,3}$$

The analysis of the emission saving by assuming 10×10^5 km distance covered by the EVs have been calculated. The energy required for the EV to cover 10×10^5 km is given:

$$E_{k=1,2,3} = EC_{k=1,2,3} * 10 \times 10^5 \tag{6}$$

The percentage energy requirement more required by different EVs compared to the other is given by

$$Emore_{k=1,2} = \frac{EC_{k+1} - EC_k}{EC_k} \tag{7}$$

The annual CO_2 emission from different EVs are given by the following equation:

$$ACE_{k=1,2,3} = CE_{k=1,2,3} * 10 \times 10^5 \tag{8}$$

The total CO_2 emission saving (when charged from FPVS instead of conventional grid) for different types of EVs are given by the following equation:

$$TCE_{k=1,2,3} = \frac{(CEcp - CEpv) \times E_{k=1,2,3}}{1000} \quad (9)$$

Finally, the CO_2 emission saving when FSPVS is used to charge the different types of EVs are given by

$$ES_{k=1,2,3} = TCE_{k=1,2,3} - ACE_{k=1,2,3} \quad (10)$$

2.6. Analysis in Evaporation

FSPVS are self-regulating and the FSPVS may increase their producing efficiency by 11% over land-based PV systems as temperatures rise [26]. In addition to limiting water evaporation, the floating solar panels also block excessive sunlight, avoiding an algal growth in the process. Layout of the PV site is shown in Figure 7.

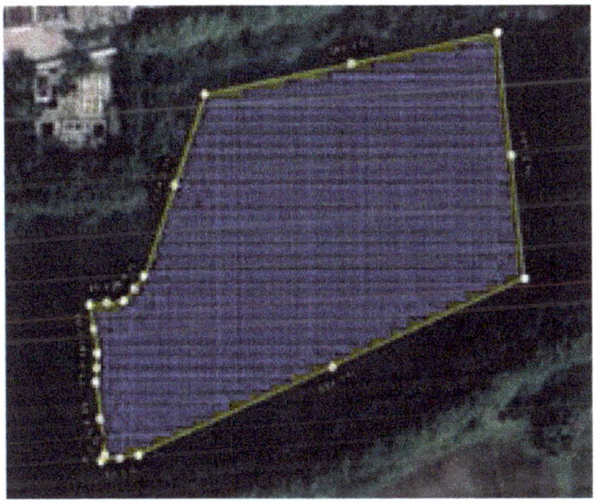

Figure 7. Layout of the PV site.

The perimeter proposed PV plant is 53 × 10^4 m while the area of the PV plant is 0.15 × 10^{11} m^2. The overall evaporation/ water saving will be 37,125 kL/annum [31], lowering the plant's specific water usage.

3. Results

3.1. Floating Plant Analysis

Block diagram of the floating SPV plant is shown in Figure 8. The GHI, POA irradiation and shaded irradiation from the floating plant obtained by month is given below in Figure 9a. One can see that the radiation increased from January to March. It then decreased from March to December. About 200.2 kWh of GHI was present during the month of March. The least amount of GHI can be seen in the month of November where the GHI was 130.3 kWh/m^2. A similar trend is seen for POA and shaded radiation. It can be seen that the POA was highest in March, at about 206.6 kWh/m^2, while it was lowest during June and July with about 145.4 kWh/m^2. The total energy extracted from the grid and the nameplate energy is given below in Figure 9b. The nameplate kWh is the energy that should be generated under the ideal case. There are a many losses associated with a floating PV system. such as AC system losses, inverter losses, charging losses, wiring losses, mismatch losses, temperature losses, shading losses, reflection losses, soiling losses and irradiance losses. The highest loss was due to reflection losses of about 3.2%. There were also a significant number of losses due to soiling. There were also significant losses due to the SMA Sunny Tripower 24,000 TL-US inverter. The inverter has an approximate efficiency of 98% and thus there was about 2% loss due to the inverter. Figure 9c shows the distribution of different losses and Figure 9d describes the performance ratio over different months of a year.

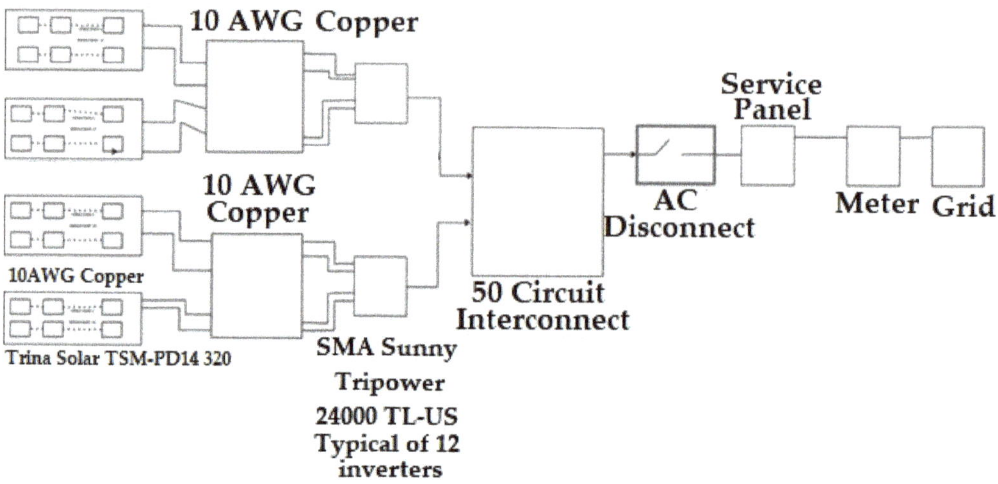

Figure 8. Floating plant block diagram.

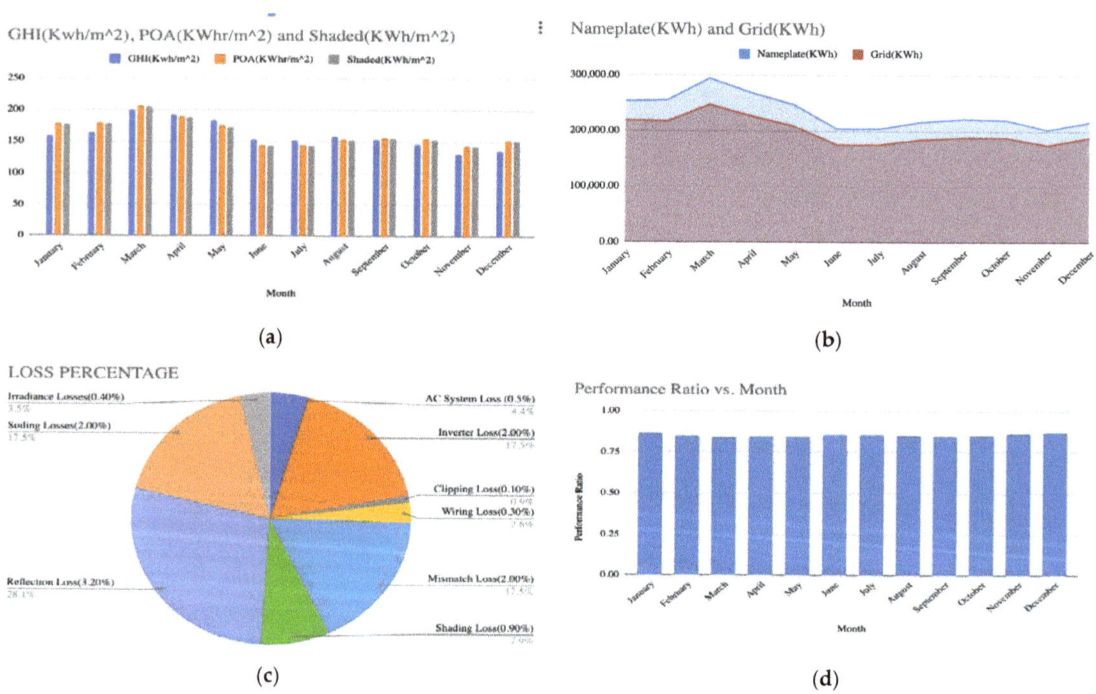

Figure 9. (**a**) GHI, POA and shaded radiation; (**b**) nameplate energy and grid energy; (**c**) loss % age; and (**d**) performance ratio.

The total energy provided to the grid was 2387.89 MWh while the nameplate energy was 2800.03 MWh. This is because there were various losses. Firstly, there were losses due to output at irradiance level which led to a decrease in the energy to grid of about 0.4%. Then there were losses due to cell temperature which led to losses of up to −9.2%. There was a further decrease of up to −2.9% due to mismatch losses. Furthermore, when the optimal DC output is considered, there was a further decrease leading to 2450.03 MWh of energy. When the constrained DC output is considered, there was a further reduction in the actual energy provided to the grid of 0.1% leading to 2447.69 MWh. Considering the inverter output, there was a further decrease in energy of up to 2.0%. Thus, the final energy provided by the floating PV system to the grid was 2387.89 MWh.

The actual energy generated was 2387.89 MWh while the energy generated under the ideal case was 2800.03 MWh. Thus, the performance ratio of the floating PV system was 0.8528 and the performance ratio was 0.8528. The calculation of the performance ratio of the PV system was also calculated for every month and it is given in Figure 9d. The performance ratio of the PV system was also calculated for every month and is given in Figure 9d. The above graph in Figure 9d shows the performance ratio by each month. The highest performance ratio was given during the month of December with a performance ratio of 0.87. It can be seen that the performance ratio was high at the beginning of the year with a performance ratio of 0.86 in January. The performance ratio slowly decreased until March where the performance ratio was 0.83. It increased in April where the performance ratio was 0.84 but then it again decreased during the month of May where the performance ratio was 0.84. It then increased after the month of May gradually until December where it reached 0.87.

3.2. Charging of EVs from FSPVS

In this work, three types of electric vehicles were considered: electric scooters, electric bikes and electric cars. The annual CO_2 emission savings are tabulated in Table 3.

Table 3. Annual CO_2 emission savings for different types of EVs when charged from FSPVS.

Parameters		Value
Annual distance covered	D_1	302.3×10^6 km
	D_2	77.8×10^6 km
	D_3	15.8×10^6 km
Percentage km covered more	$Dmore_1$	289
	$Dmore_2$	392
CO_2 emissions (g/psngr/km)	CE_1	16.1 g/psngr/km
	CE_2	29.8 g/psngr/km
	CE_3	92.4 g/psngr/km
Annual CO_2 emissions	ACE_1	4866.5 Ton/psngr
	ACE_2	2317.9 Ton/psngr
	ACE_3	1461.2 Ton/psngr
CO_2 emissions from plants	Coal based plants	1000 g/kWh
	PV based plant	40 g/kWh
Savings in CO_2 emissions		960 g/unit
Total CO_2 emission savings	TCE	2292.38 Ton
Annual emission savings when EV charged from FPVS	Electric bike	−2574.1 Ton
	Electric scooter	−25.5 Ton
	Electric car	831.2 Ton
Net Zero CO_2 emission time period (months)	Electric bike	25.5
	Electric scooter	12.1
	Electric car	7.7

The annual CO_2 emission was also measured by assuming a fixed coverage distance of 10×10^5 km. The annual CO_2 emission of bike, scooter, and car was obtained by the equations mentioned above. Finally, the emission saving when PV is used along with electric bike, scooter and car was calculated and compared. Table 4 presents the CO_2 emission parameters assuming constant distance coverage.

Table 4. CO_2 emission parameters assuming constant distance coverage.

Parameters		Value
Energy required	E_1	7.9 Wh/km
	E_2	30.7 Wh/km
	E_3	151 Wh/km
Percentage energy more required	$Emore_1$	289
	$Emore_2$	392
Annual CO_2 emissions	ACE_1	161 Ton/psngr/km
	ACE_2	298 Ton/psngr/km
	ACE_3	924 Ton/psngr/km
Total CO_2 emission saving when EV charged from FPVS instead of conventional grid	TCE_1	75.84 Ton/psngr/km
	TCE_2	294.72 Ton/psngr/km
	TCE_3	1449.6 Ton/psngr/km
Emission savings when FPVS is used to charge EV	Electric bike	−85.16 Ton
	Electric scooter	−3.28 Ton
	Electric car	525.6 Ton

It can be seen that the CO_2 emission savings was positive in the case of the E-car, but it turned out to be negative in the case of the E-bike and E-scooter. According to the research, when charged by floating PV, the net CO_2 emission for an e-bike, e-scooter, and e-car, respectively, was zero in just 25.5 months, 12.1 months, and 7.7 months. All three types of electrical cars were zero emission vehicles after the aforementioned time periods. The analysis also took into account the predetermined distance (10×10^5 km) that each of the three types of vehicles must travel in order to charge. When using a floating PV system to charge electric vehicles, carbon emissions were reported to be positively reduced for electric vehicles but not for electric scooters or cycles. In such situation, it was estimated that using an electric bike, scooter, or automobile would result in CO_2 emission reductions of −85.16 Ton, −3.28 Ton, and 525.6 Ton, respectively. For countries that rely on renewable energy generation, analysis reveals that all three types of electric vehicles are successful in lowering CO_2 emissions, but when a set distance travelled is taken into account, only electric automobiles reduce carbon emissions. The electric car is the best option for reducing CO_2 emissions out of all the electric vehicles that have been taken into consideration.

4. Conclusions

In this article the performance of a 1.5 MWP floating PV system using Helioscope software has been simulated and performance of the same has been analyzed to charge different types of EVs, such as the E-bike, E-scooter and E-car. The annual energy generation from the 1.5 MWp floating PV plant is of 2387.89 MWh. The analysis shows that the charging of EV using a floating PV system is more efficient from the point of view of CO_2 emission. The findings indicate that an E-bike, E-scooter, or E-car will be a net zero CO_2 emission vehicle in 25.5 months, 12.1 months, and 7.7 months, respectively, when charged from floating PV plant. When the floating PV system is used to charge various types of electric vehicles for a fixed distance coverage of 10×10^5 km, it has been observed that the CO_2 emission savings for the E-bike, E-scooter, and E-car are −85.16 Ton, −3.28 Ton, and 525.6 Ton, respectively. According to the findings of the analysis, all three types of electric vehicles are effective in reducing CO_2 emissions for nations that rely on renewable energy generation; however, when considered for a fixed distance covered, only electric cars save

carbon emissions. It can be seen that, among the various types of electric vehicles that have been considered, the E-car is the most effective choice in terms of the reduction in CO_2 emission. Additionally, 37,125 kL of water will be saved annually from evaporation owe to the installation of the floating PV plant.

Author Contributions: Data curation, methodology, visualization, writing—original draft preparation, A.R.; Conceptualization, software, formal analysis, investigation, validation, S.C.; writing—review and editing, supervision, D.E.; writing—review and editing, resources, data curation, supervision, funding acquisition, H.K.; resources, data curation, K.M.A., resources, data curation, methodology, visualization, N.C.G.; supervision, project administration, E.B.A. All authors have read and agreed to the published version of the manuscript.

Funding: This research received no external funding.

Institutional Review Board Statement: Not applicable.

Informed Consent Statement: Not applicable.

Data Availability Statement: The data sources employed for analysis are presented in the text.

Conflicts of Interest: The authors declare no conflict of interest.

References

1. Ellison, E.; Baker, L.; Wilson, A. IPCC Special Report Meeting: Climate Change Around the Globe. *Weather* **2020**, *75*, 293–294. [CrossRef]
2. Shenoy, S.; Gorinevsky, D.; Trenberth, K.E.; Chu, S. Trends of extreme US weather events in the changing climate. *Proc. Natl. Acad. Sci. USA* **2022**, *119*, e2207536119. [CrossRef] [PubMed]
3. Biros, C.; Rossi, C.; Talbot, A. Translating the International Panel on climate change reports: Standardisation of terminology in synthesis reports from 1990 to 2014. *Perspectives* **2020**, *29*, 231–244. [CrossRef]
4. Kobzar, O.; Melnyk, V.; Boon, E.K.; Derykolenko, O.; Kharchenko, M.; Karintseva, O. Environmental determinants of energy-efficient transformation of national economies for sustainable development. *Int. J. Glob. Energy Issues* **2021**, *43*, 262. [CrossRef]
5. Kumar, M.; Niyaz, H.M.; Gupta, R. Challenges and opportunities towards the development of floating photovoltaic systems. *Sol. Energy Mater. Sol. Cells* **2021**, *233*, 111408. [CrossRef]
6. Haugwitz, F.; Advisor, S. Apricum Floating Solar PV Gains Global Momentum. Available online: https://www.pv-magazine.com/2020/09/22/floating-solar-pv-gains-global-momentum/ (accessed on 25 November 2022).
7. Kumar, M.; Kumar, A.; Gupta, R. Comparative degradation analysis of different photovoltaic technologies on experimentally simulated water bodies and estimation of evaporation loss reduction. *Prog. Photovolt. Res. Appl.* **2020**, *29*, 357–378. [CrossRef]
8. Available online: https://www.globenewswire.com/en/news-release/2022/02/24/2391102/0/en/Global-Floating-Solar-Panels-Market-to-Reach-4-8-Thousand-MW-by-the-Year-2026.htm (accessed on 25 November 2022).
9. Floating Solar Market Size to Hit US$ 10.09 Billion by 2030. Available online: https://www.precedenceresearch.com/floating-solar-market (accessed on 25 November 2022).
10. Sahu, A.; Yadav, N.; Sudhakar, K. Floating photovoltaic power plant: A review. *Renew. Sustain. Energy Rev.* **2016**, *66*, 815–824. [CrossRef]
11. Goswami, A.; Sadhu, P.K. Degradation analysis and the impacts on feasibility study of floating solar photovoltaic systems. *Sustain. Energy Grids Netw.* **2021**, *26*, 100425. [CrossRef]
12. Oliveira-Pinto, S.; Stokkermans, J. Assessment of the potential of different floating solar technologies–Overview and analysis of different case studies. *Energy Convers. Manag.* **2020**, *211*, 112747. [CrossRef]
13. Taye, B.Z.; Nebey, A.H.; Workineh, T.G. Design of floating solar PV system for typical household on Debre Mariam Island. *Cogent Eng.* **2020**, *7*, 1829275. [CrossRef]
14. Goswami, A.; Sadhu, P.; Goswami, U.; Sadhu, P.K. Floating solar power plant for sustainable development: A techno-economic analysis. *Environ. Prog. Sustain. Energy* **2019**, *38*, e13268. [CrossRef]
15. Costa, L.C.A.; Silva, G.D.P. Save water and energy: A techno-economic analysis of a floating solar photovoltaic system to power a water integration project in the Brazilian semiarid. *Int. J. Energy Res.* **2021**, *45*, 17924–17941. [CrossRef]
16. Ravichandran, N.; Panneerselvam, B. Performance analysis of a floating photovoltaic covering system in an Indian reservoir. *Clean Energy* **2021**, *5*, 208–228. [CrossRef]
17. Agrawal, K.K.; Jha, S.K.; Mittal, R.K.; Vashishtha, S. Assessment of floating solar PV (FSPV) potential and water conservation: Case study on Rajghat Dam in Uttar Pradesh, India. *Energy Sustain. Dev.* **2022**, *66*, 287–295. [CrossRef]
18. Available online: https://www.oecd.org/publications/global-ev-outlook-2020-d394399e-en.htm (accessed on 25 August 2022).
19. Prospects for Electric Vehicle Deployment–Global EV Outlook 2021–Analysis. Available online: https://www.iea.org/reports/global-ev-outlook-2021/prospects-for-electric-vehicle-deployment (accessed on 25 November 2022).
20. Available online: http://www.hydrogen.energy.gov/pdfs/14006_cradle_to_grave_analysis (accessed on 25 September 2022).

21. Michalek, J.J.; Chester, M.; Jaramillo, P.; Samaras, C.; Shiau, C.-S.N.; Lave, L.B. Valuation of plug-in vehicle life-cycle air emissions and oil displacement benefits. *Proc. Natl. Acad. Sci. USA* **2011**, *108*, 16554–16558. [CrossRef] [PubMed]
22. Shen, W.X.; Zhang, B.; Zhang, Y.F.; Wang, X.C.; Lu, Q.; Wang, C. Research on Life Cycle Energy Consumption and Environmental Emissions of Light-Duty Battery Electric Vehicles. *Mater. Sci. Forum* **2015**, *814*, 447–457. [CrossRef]
23. Rajneesh, K.; Ananya, R.; Rohan, B. Prediction of Global Horizontal Radiation in Vellore using Clearness Index Model. *Int. J. Eng. Technol.* **2017**, *9*, 489493.
24. Rahman, M.; Hasanuzzaman, M.; Rahim, N. Effects of various parameters on PV-module power and efficiency. *Energy Convers. Manag.* **2015**, *103*, 348–358. [CrossRef]
25. Available online: http://www.reuters.com/business/energy/Singapore-unveils-one-worlds-biggest-floating-solar-panel-farms-2021-07-14/ (accessed on 29 September 2022).
26. Suh, J.; Jang, Y.; Choi, Y. Comparison of Electric Power Output Observed and Estimated from Floating Photovoltaic Systems: A Case Study on the Hapcheon Dam, Korea. *Sustainability* **2019**, *12*, 276. [CrossRef]
27. Umoette, A.T.; Ubom, E.A.; Festus, M.U. Design of Stand Alone Floating PV System for Ibeno Health Centre. *Sci. J. Energy Eng.* **2016**, *4*, 56. [CrossRef]
28. Hsu, S.A. Correction of Land-Based Wind Data for Offshore Applications: A Further Evaluation. *J. Phys. Oceanogr.* **1986**, *16*, 390–394. [CrossRef]
29. McCafferty, D.J.; Gilbert, C.; Thierry, A.-M.; Currie, J.; Le Maho, Y.; Ancel, A. Emperor penguin body surfaces cool below air temperature. *Biol. Lett.* **2013**, *9*, 20121192. [CrossRef] [PubMed]
30. The Environmental Impact of Today's Transport Types. Available online: https://tnmt.com/infographics/carbon-emissions-by-transport-type/ (accessed on 25 November 2022).
31. Kumar, N.M.; Chakraborty, S.; Yadav, S.K.; Singh, J.; Chopra, S.S. Advancing simulation tools specific to floating solar photovoltaic systems–Comparative analysis of field-measured and simulated energy performance. *Sustain. Energy Technol. Assess.* **2022**, *52*, 102168. [CrossRef]

MDPI
St. Alban-Anlage 66
4052 Basel
Switzerland
www.mdpi.com

Applied Sciences Editorial Office
E-mail: applsci@mdpi.com
www.mdpi.com/journal/applsci

www.ingramcontent.com/pod-product-compliance
Lightning Source LLC
Chambersburg PA
CBHW061629210926
51850CB00009B/2484
* 9 7 8 3 0 3 6 5 8 6 2 8 1 *